Making Better Sense of the World

Making Better Sense of the World

Bruce S C Robertson

Copyright © 2019 by Bruce S C Robertson.

Library of Congress Control Number:		2019912160
ISBN:	Hardcover	978-1-5434-9570-6
	Softcover	978-1-5434-9569-0
	eBook	978-1-5434-9568-3

All rights reserved. No part of this book may be reproduced or transmitted in any form or by any means, electronic or mechanical, including photocopying, recording, or by any information storage and retrieval system, without permission in writing from the copyright owner.

Print information available on the last page.

Rev. date: 09/26/2019

To order additional copies of this book, contact:
Xlibris
0-800-443-678
www.Xlibris.co.nz
Orders@Xlibris.co.nz
800430

The front cover depicts a variation of a Mandelbrot set. It was generated on a computer from a very simple mathematical function. The complexity that can emanate from such simplicity resonates with the complexity of ideas that can be generated from a few simple assumptions such as those described in this book.

Contents

Introduction..vii
Chapter 1   Philosophy is a paradigm.........................1
Chapter 2   Pattern Identification...........................7
Chapter 3   Anchor points and abstract systems..............15
Chapter 4   Creating patterns...............................25
Chapter 5   Interacting with the world......................29
Chapter 6   Communication...................................37
Chapter 7   Decision Making.................................45
Chapter 8   Logical Systems and Logical Processors..........53
Chapter 9   Pyramid of Patterns.............................71
Chapter 10 Decision Making.................................83
Chapter 11 Making better sense of the world................89

Glossary...117
Addendum...121

Introduction

I am not one to follow a recipe. I use recipes as a source of ideas and suggestions rather than as a set of instructions to be followed. My culinary skills probably leave much to be desired, but that is more to do with my lack of interest in cooking than in my inability to follow instructions.

When I go kayaking up an estuarine inlet I like to follow the creek until I can go no further, I either encounter an impassable waterfall or the creek narrows until it is no more than a drainage ditch.

I have taken the same approach to philosophy. I use the writings of philosophers past as a source of ideas and suggestions and possibilities rather than as something to be revered and respected. But unlike cooking I am very interested in philosophy and over many years I have honed my skills and knowledge in the field and have found many things that are most interesting. And as in kayaking I have tried to go as far as I possibly can to find the source.

A few years ago I completed 'The Pattern Paradigm', a book that laid out some theories for the basis of knowledge, both empirical and theoretical, real and abstract. It was the first of a trilogy (*The Pattern Paradigm* trilogy). This book is a sequel to the first and the second of the trilogy. It builds upon the epistemological foundations of The Pattern Paradigm. It takes the ideas forward and while adopting the same degree of rigour and making as few assumptions as possible it uncovers some interesting ideas and leads to the basis for a simple and comprehensive philosophy.

Something that I have realised along the way is that philosophy is personal. There is no absolute and undeniable philosophy that is objectively and universally true and for which all that is required is for this truth to be excavated and identified and then all will be understood. It doesn't work that way at all. Philosophy is personal, for if it is not believed, it is nothing.

I believe the philosophy described in this book. I do not claim that it is the only one possible, but I believe it to be the best.

I have no doubt that everyone believes that their philosophy is the best, for otherwise they would not believe it. But few, if any, can write out a philosophy that is as clear, concise and comprehensive and with as few presumptions as the one described in this book.

Philosophy is also about communication; and the medium for communication is language or more specifically in the case of this book, the English language. So that is why I'm writing this book.

It has taken me longer than I originally intended to write this book. For as I wrote the book, new ideas would come along and I would want to pursue them as they could be relevant to the book. Also I find that exploring new ideas and making notes about them is more enjoyable than actually writing them all out in a logical order and in a form that can be understood by other people. But the writing has also been a constructive exercise as it has forced me to examine the relationship between various ideas and to fill in gaps that I might encounter.

The philosophy described in this book breaks from the traditions of contemporary philosophy. It doesn't seek to build upon the writings of philosophers past, instead it seeks to go deeper and build upon a few simple assumptions. Also its methods are somewhat different. It is not an analysis of normative assumptions but is instead a synthesis from a few core ideas and builds from them to arrive at a description of the normative world. In starting from the elementary concepts it works its way outwards in a helical fashion; on first introducing a topic it describes it very simply and then later on when other concepts have been introduced it revisits the topic and describes it in a little more detail.

If I were to put it into the context of the canon of philosophical writings, I would say that it is most compatible with and perhaps the continuation of the work of David Hume. But it is also an original approach to philosophy in that it delves beneath the level of language and into the logic of thought processes that lie beneath, and that perhaps is its biggest difference from other philosophical approaches.

So I am starting this book, not with a description of the pattern paradigm philosophy but with a discussion of how differing philosophies with differing methodologies can be meaningfully compared and evaluated.

This book is about making sense of the world. It is not about the way the world 'should' be, it is not even about the way the world 'is'. Instead it is a description of the way the world appears and about the best way to view the world. In order to make sense of the world, it is necessary to go fairly deep into the processes of creating a model of the world.

Chapter 1

Philosophy is a paradigm

Philosophy is all about making sense of the world and how one can make sense of the world is what this what this book is all about.

Making sense of the world is a personal thing. And philosophy, despite what many would claim, is also a personal thing. There can be no definitive proof that one philosophy is 'better' than another; instead it is a matter of personal choice. Which particular philosophy a person chooses will depend upon their personal judgement and will be based upon factors such as environment, experiences and knowledge. It is not like a science such as physics where the proof of the theories can be demonstrated through technological advances.

However there are some aspects of making sense of the world which would seem to be essential to any philosophy. Foremost of these is the idea that the human brain creates a model of the world or, if you like, a picture of the world. This model is distinct from the world itself. Kant labelled the model and the real world as the 'phenomenon' and the 'noumenon' respectively. The phenomenon being that which is perceived and the noumenon being the thing that actually exists. And while everyone creates a model of the world, each model is slightly different and reflects the different qualities of the person's brain (just as people's faces are slightly different, so too are their brains), their differing environments, cultures and experiences.

In order to create a model of the world, certain assumptions need to be made, for without any assumptions there could be no starting point and no foundation upon which to build a model. And these assumptions can be considered to identify the particular philosophy and distinguish it from others.

From a personal perspective, the model that one holds is one's entire world. There is nothing beyond it. However from a philosophical perspective, models that are sufficiently different from others can be considered to be paradigms. It would be expected that each viable paradigm would be capable of describing the world and also to be internally self-consistent.

Each paradigm would have its own assumptions and perhaps also its own method for making inferences about the world. So, for example, in a Christian paradigm the claim that 'God made the world' could be considered to be an essential assumption or foundation for that paradigm. The assumption that 'Truth is a property of statements' could be a foundation for another paradigm.

Whenever an assumption is made, whether implicit or explicit, the implications that follow from it are necessarily part of a paradigm. An example of an implicit assumption is that of the question: 'What should a person do in a particular situation?' It is implicit in that question that there is something that the person 'should' do. This is an implicit assumption and everything that follows from that question is a paradigm.

Philosophy is all about paradigms. For there is no comprehensive philosophy that is without assumptions and hence every comprehensive philosophy must constitute a paradigm.

The difference between 'truth' and a paradigm is that a paradigm is acknowledged as being capable of shifting its ideas and theories whereas 'truth' generally is not. And while 'truth' might seem preferable to a paradigm, it requires a naive view of the world, whereas acknowledging that knowledge of the world is a model and consists of a paradigm, does not.

Since there are a number of paradigms, each of which has its own assumptions and internal processes of inference, it cannot be claimed that there is just one true paradigm and that the rest are false; at least not from an objective philosophical perspective. However there may be some criteria by which different philosophical paradigms can be compared and then a selection made for the one that fits one's personal perspective best.

These criteria might include:

1. Simplicity. A useful paradigm will be simple, and simplicity is preferable to complexity.

2. Accuracy. A paradigm that accurately fits one's knowledge of the world is preferable to one that does not.

3. Comprehensiveness. A philosophical paradigm that covers the full range of knowledge is preferable to one with only a limited range.

4. Internal self-consistency. A philosophical paradigm which allows for a smooth transition from one topic to another is preferable to one that is compartmentalised.

This list of criteria is not intended to be comprehensive, it is just an indication of how different paradigms can be compared.

But if there is no definitive process for evaluating the merits of a philosophical paradigm and the selection made depends upon various personal criteria, one could ask: Does it matter which paradigm one chooses? Or perhaps: Does philosophy matter at all?

Well, yes it does. A person's philosophy underpins and influences every decision they make. And over a lifetime this will total a lot of decisions that have been influenced and a lot of selections of paths to follow.

It would seem reasonable to claim that a person with a 'good' philosophy (by 'good' I mean one that conforms to the criteria cited above and one that suits a person's particular circumstances) will have a better likelihood (other factors being equal) of a successful and happy life than a person with a not-so-good one.

The best philosophical paradigms are based upon logic. And here I am using the term 'logic' in the general sense of using axioms (whether explicit or implicit) to draw inferences. The axioms can be considered to be assumptions. They are hypothesised out of nothing. The test for their efficacy is in the usefulness of their inferences. I am emphasising the use of this logic as essential to philosophy for if there is no logic behind the inferences then they can only be considered as random, chaotic and ultimately meaningless.

But logic cannot be applied or used without a logical processor: a logical machine that can interpret the axioms to generate a conclusion. Typically this can either be a brain or a computer.

In the case of a computer that has been programmed to follow particular logical procedures, the logical processes are known. However, most often in logic, and particularly in classical philosophy, it is a brain that is the logical processor. And since the logical workings of the brain are hidden one can only infer from its inputs and outputs what logical processes it actually follows.

My earlier book: 'The Pattern Paradigm' explored the likely logical processes of the brain and how they fitted together. This book takes this a step further to the application of those logical processes and how they can be used to make sense of the world.

Since each brain is different from every other brain the actual processes and actual inputs and outputs of a brain will be slightly different. The actual model of the world that each brain creates will be different from every other. Nevertheless, it may be that there is considerable commonality of the logical processes that are used by the brain. So this book focuses mainly on the processes used in making a model of the world and only making the broadest of inferences based upon these processes and only using the most common of general knowledge as a starting point.

As mentioned earlier, every philosophy is personal and the philosophy described in this book is no different. But by focusing on the logical methods rather than on their conclusions, I am hoping that a commonality can be found that is applicable to all people.

Every person has their own model of the world. In this book I want to explore how that model is created. I'm interested in the logical processes that are necessary for the creation of a model of the world. And having identified those processes I want to use them to describe the world that is commensurate with those processes.

Philosophy, like any academic discipline, begins with an examination of normative facts and categorisation of obvious properties, in other words there is an analysis of the available data. However analysis can only go so far, for it has to assume that the normative data is essentially accurate and this may not always be the case. For example, many years ago it was generally accepted that the world was flat and easy to make conclusion based on limited data, but as more data became available it is seen that this presumption is less than accurate and a more rounded world makes better sense. So it is also with philosophy, what appear to be facts cannot necessarily be relied upon.

A more powerful processors and analysis is synthesis, a putting together or a few simple ideas to see if they can be combined to create a model which can generate the known data. This may not exactly fit the known facts as those facts may be less than accurate. And if it does not match those facts perfectly then this need not be a failing of the synthesis.

The genesis of this book began many years ago with an analysis of the relevant data and a search the deepest theories or patterns that can describe the data. It also incorporated a refusal to accept at face value is assumptions made by other philosophers and other philosophical paradigms.

This book, along with its prequel 'The Pattern Paradigm' is a result of those endeavours. A simple synthesis which fits the data and reconstitute the facts. Some of its inferences and conclusions may be somewhat askew from popular beliefs but that is akin to claiming that the world is round to people who believe that the world is flat. This philosophical paradigm may be referred to henceforth as 'TPP'.

The main text of this book is a synthesis. It combines a few powerful ideas to arrive at some interesting conclusions.

But first we need to start at the beginning.

Chapter 2

Pattern Identification

In the beginning there was no cognition, no concepts, no understanding and no model of the world. So how is it possible that a state such as that could transition into one in which there is cognition, concepts, understanding and a model of the world? What logical process, or processes, could enable this to occur?

From our modern perspective of the world, the Earth was created some four billion years ago. And some while later the beginnings of life appeared. Early life, akin to modern plants, interacted with the world in a mechanical way without any concepts or cognition or a model of the world. All that was required was that it respond to stimuli. So, for example, a flower, on receiving heat from the sun, might open its petals to take full advantage of the sunlight. But it does so in a mechanical way, like a thermostat with no concept of 'heat, 'sun' or 'petal' being necessary or required.

Mechanical responses, through evolution, can evolve into more complex mechanical responses but they cannot create a model of the world using that same simplistic logic of stimuli and response. It can only lead to more intricate and complex responses to stimuli.

Fast forward to the present and modern human life displays evidence of a comprehensive model of the world including concepts of space, time and matter. This could only have occurred through a qualitative change in the logic of how a lifeform interacts with the world. But what change could that be?

The situation is analogous to that of seeing a tortoise at one side of a deep and seemingly impassable chasm and some while later seeing the same tortoise at the other side of the chasm and realising that somehow it must have travelled from one side of the chasm to the other. The fact that it was once at one side of the chasm and subsequently at the other side seems indubitable, yet there does not appear to be any obvious means by which the tortoise could have traversed the chasm. The task of the philosopher is to explore the possible ways that the tortoise could have traversed the chasm.

Consider the situation of an organism that exists at the very dawn of animal life. We can hypothesise that it has a logical processor in the form of a primitive brain which is linked to some senses that can detect data from the world and one or more motor outputs that allow the organism to interact with the world.

The question comes down to this: How can sense-data be processed so that a concept of what lies behind the data can be conceived?

It needs to be emphasised that in this stage there are no concepts of the real world nor a concept of self and not even any concept of senses. The classical philosophical claim of the form:"There is a tree in front of me. I can see it, touch it and smell it and hence I can identify it as a tree" is, in this instance, entirely ineffectual. For our primitive organism has no concept of a tree with which to match the data from its senses. It does not even have any concept of seeing or touching. It may receive data from its light sensors and audio senses but it would have no knowledge that the data corresponded to seeing or hearing.

I contend that there is only one possible way that data can be processed with no prior knowledge as to what the data might correspond to and which has the possibility of leading to concepts of the real world. And that is through the logical process of pattern identification. Details of how this pattern identification process works is described in my earlier book: *"The Pattern Paradigm"*; however I will reiterate the main points here.

The process of pattern identification can be modelled using mathematics and algorithms. The data that is incident upon the senses can be modelled by a sequence of integers. Computers store data for pictures and sound as strings of binary integers so clearly visual and auditory data can be modelled by integers. And in the same way it is reasonable to infer that this can also apply to the other senses of touch, taste and smell.

The logical processes of the primitive brain can be modelled using algorithms (a sequence of logical instructions). It is the task of the algorithm to find a pattern that best fits the data. The algorithm would have to incorporate a form of trial and error using various templates which are then tested to determine which, if any, works best.

The outcome of this logical process would be the details of the pattern that is the best fit to the data.

The main criteria that are used to identify the 'best pattern' are:

1. Simplicity, the degree of data compression that can be achieved.
2. Accuracy, how well the pattern can reproduce the original data.

Both of these criteria can be evaluated fairly straightforwardly using logical algorithms. The relative weight given to these two criteria can be set by parameters and a final evaluation for each trialled template can be made.

So, as a trivial example, if the input data were a long string of 1000 "1"s: "1 1 1 1 1 1 1 1 ..." then a pattern that might be found would be "1000 '1's". Note that:

1. This is a compression of the data. The original data consists of 1000 symbols, and the pattern that was found consists of a mere 8.

2. The pattern, when correctly interpreted, is capable of accurately recreating the original data.

3. The pattern actually contains information that was not directly available in the original data string. (In this instance it is that there are 1000 '1's.)

4. The pattern can be obtained without any prior knowledge or concepts about the data.

5. The pattern can be projected beyond the domain of the original data.

There are other points to be noted regarding the pattern identification process:

1. The logic of the pattern identification process underpins all knowledge of the world around us.

2. Various parameters need to be set, perhaps arbitrarily, in order for the algorithm to function. These will include: how much data to analyse; how long to spend searching for a best pattern.

3. The pattern identifying logic requires input of templates which it can use to search for a possible pattern. These templates can be randomly generated. However the system can only search the patterns based upon this limited input.

4. When a best pattern has been selected and it is identified as being a particularly good pattern it can be labelled as a "good pattern" to distinguish it from other best patterns that are perhaps not quite so good.

5. If data is subsequently received that does not tally with an existing pattern of the same domain then it may be expedient to rerun the pattern identifying process for that domain.

6. A good pattern will have the capability of reproducing the original data, at least approximately.

7. It is likely that even the best pattern will only have an approximate fit to the data.

Note that in this example, the data is "111111………." and the pattern is "1000 1's".

It is through this pattern identification process that the seeds of perception and cognition are generated.

Some of the early patterns formed in this way will be those that correspond to the patterns that are now normatively labelled as 'time' and 'space', as these would seem fundamental to an understanding of the world.

Kant suggested that such concepts were 'a priori' but he didn't elaborate upon where or how such 'a priori' concepts originated from. The pattern identification process described here both explains it and simplifies it.

Following the logic of the system, it is quite possible that when a sufficient number of patterns have been identified from the raw sense data and stored, the pattern identification process can be run recursively using the stored patterns as the input data. In this way higher order patterns can be identified and stored.

And when a sufficient quantity of these second order patterns have been found, they can be used as input for the pattern identification process to produce third order patterns and so on.

Each successive level of patterns would necessarily be smaller than the one below it. So one can think of the stored patterns taking on a pyramid type structure with many patterns produced from the raw sense data at the lowest level and with each successive layer being a little smaller.

But how does this pyramid of patterns fit in with the necessities of life?

The pyramid of patterns that is created provides the basis for concepts and ideas and thinking. Without a pyramid of patterns, there are no concepts. An organism without a pyramid of patterns may be able to interact with the world in a mechanical way and even be able to refine the way it interacts with the world in a way that could be described as a form of crude learning; but it would have no concept of the world nor of its own existence.

Some of the first patterns that an organism's brain will generate are likely to be those that we associate with 'time' and 'space' as these would seem fundamental to an understanding of the world.

The patterns that are created and stored in a brain constitute 'beliefs'. There are no false beliefs; there are only those beliefs that are held to be the best possible, while those that are not, are discarded.

A situation may arise that as more data is gathered that relate to a particular domain and the pattern that had originally been identified for that domain no longer seems to be a good fit, then it would be expedient to rerun the pattern identification process for all the data in that domain and to arrive at a new and hopefully better pattern. If this happens at a fairly deep level in the pyramid of patterns then the reallocation of that pattern will have a ripple effect as its influence proceeds up the pyramid of patterns. This is what could be called a 'paradigm shift'.

A classic example of a paradigms shift occurred when Einstein published his 1905 paper on special relativity when it became apparent that the normative patterns for space and time that most people had identified did not fit the new data and a new pattern was identified which combines elements of time and space into space-time.

A more common paradigm shift, but of a more minor nature, is of the type when one receives some additional data with regard to a trusted friend that does not fit with that pattern and instead a new pattern is generated for which the friend is not so trustworthy. This paradigm shift can have a ripple effect throughout one's pyramid of patterns and also in one's interactions with the friend.

The patterns that are stored or likely to be labelled to allow for easy reference and usage. Some of the patterns at the upper end of the pyramid of patterns (but not near the base) will have labels that can be spoken and communicated such as 'tree', 'house' and 'pineapple'.

Creating a pyramid of patterns from sense data is however only part of the process of generating a useful model of the world. The other part is using those patterns as a basis for interacting with the world to find the necessities of life such as food and shelter. It is only through this interaction with the world that the patterns can be useful.

A plant can interact with the world using simple mechanical processes of the form 'If A then do X' or perhaps 'If B and C then Z'. In contrast an animal with a brain and even only a rudimentary pyramid of patterns can operate a far greater range of choices. So this could be 'If A then compare A with the stored patterns and then choose an output from B, C, D, E, F and G'; where B, C, D, E, F, and G represent various motor outputs of which the organism is capable. While this may not seem to be a hugely significant improvement, it is the first step towards intelligent life and the creation of a useful model of the world.

It has been shown in this section how logical processes of a brain combined with sense data can generate concepts and awareness and an efficient way for an organism to interact with the world.

It is asserted that this is the only way to have a foundation for making sense of the world.

CHAPTER 3

Anchor points and abstract systems

Philosophy without an explicit method is indistinguishable from opinion. Anyone can have an opinion and it may well be deep and philosophical but without a method to link the ideas together it remains as merely an opinion.

Aristotle identified logic as the key to philosophical method, and he was right. But what is logic?

Traditional philosophy considers logic to be the manipulation of words, but it is without a detailed and explicit process for manipulating them.

Apart from the syllogism (which in any case was not used by Aristotle), logic is frustratingly undefined and it is hard to tell what is logical and what is not. As such it does not constitute what I would describe as a good method to do philosophy.

What constitutes a logical method is a series of well-defined steps that can progress from the starting point to an end point. It should be possible that these steps can be modelled by an algorithm.

The principal method of the TPP paradigm, albeit not the only one, is that of pattern identification which was described in the previous chapter where it was asserted that such a process could be modelled by an algorithm that searched for a pattern by testing various templates.

As discussed previously, philosophy necessarily requires the making of assumptions. And as also noted previously I shall be overt and explicit in the assumptions that I make. (This is in contrast to most other philosophical paradigms.)

The assumptions that are particularly important for the TPP philosophy, I shall label as 'anchor points'.

The pattern identification process as applied to sense data and which continues as a process to the top of the pyramid of patterns is a fundamental assumption for the TPP philosophy. So I shall label this assumption as an anchor point.

Also the processes of inference used throughout this book are also those of pattern identification. I shall discuss this point later on in the book.

Another anchor point for TPP relates to 'truth'. The normative terminology truth refers to what 'is' and yet there is no method for determining whether something 'is' or 'is not' other than one of personal subjective experience. In popular philosophy it would seem that truth is poorly defined and without any explicit method of distinguishing what is true and what is not true.

However as discussed in the previous chapter a brain can only create a model of the world through a process of pattern identification. Then when a pattern is identified which constitutes the best fit to the data it can be stored and believed. Some patterns are not as clearly identified as others and these can be revised at a later date or when additional relevant data is encountered. In contrast, some other patterns may be a particularly good fit to the data. These patterns can be labelled in a way to indicate that they are a particularly good fit to the data and there is no need to revisit them for further analysis. The obviously appropriate label for this is 'true'.

So within the TPP model 'truth' is a label. And I have described a process (in effect an algorithm) by which ideas about the world can be labelled as 'true'.

Not only is this process adequately rigorous but it also fits in with the normative notion of things which are deemed to be 'true'. In other words, people label ideas or beliefs to be true when they do not doubt them.

The concept of truth in other philosophical paradigms is not so rigorously specified. Typically they treat truth as a property of an idea or statement as opposed to a label. But since ideas or statements can only relate to a model of the world rather than to the world itself it is perhaps not too surprising that this results in their concept of truth being notoriously vague and ill-defined.

The idea of truth as a label highlights the fact that all we know is a model of the world rather than the world itself and this is why 'truth as a label' is an anchor point for the TPP philosophy.

Hitherto I have been discussing a logical system (pattern identification) that is linked to the senses and which can present information about some exterior 'real world'. But one can also have logical systems that are not linked to senses. And since they have no direct link to any 'real world' they can be considered to be 'abstract'.

Hume (and others) noted that there are two distinct types of knowledge: 'Facts' and 'Relations of ideas'. This distinction is followed in TPP. However I prefer to refer to them as 'real' and 'abstract', as this highlights the connection that pattern identification has to the senses and to the outside world; whereas logical systems that are not connected directly to the senses can be considered to be abstract.

The most obvious example of an abstract logical system is that of pure mathematics, but there are others.

There have been a number of suggestions for a philosophy of mathematics, though what is really required from a philosophy of mathematics is a means or model by which mathematics can be comfortably integrated with other areas of knowledge and philosophy.

The model of mathematics that would seem to best fit with TPP is that of an abstract system that is founded on abstract axioms and which generates theorems by processes that are defined by rules and algorithms.

This is discussed in more detail in my earlier book: '*The Pattern Paradigm*', but I will go over the main points here.

An abstract system consists of axioms which define the system, including elements and processes of logical inference, and theorems which are the output of the abstract system.

The axioms and rules suggested for an abstract system can be pretty much anything, but if they are to define a well-tempered abstract system then they must fulfil a number of requirements.

1. The system must be able to create theorems.
2. The axioms must be sufficiently compatible with each other that a unique abstract machine can be constructed which incorporates all those axioms and which can (at least in theory) generate the theorems.
3. The system, at its deepest level, must follow simple rules which operate on symbols.

One might think that another constraint on such systems would be that they do not generate theorems which contradict each other. However the theorems, at their most fundamental level, consist of nothing more than strings of symbols which have no meaning and so there can be no such thing as a contradiction.

In these respects it is very similar to what is commonly called a 'formal system', however I am calling it an abstract system to highlight its abstract nature and to contrast it with a 'real' system. A criticism of the formal system is that its axioms are somewhat arbitrary. An abstract system embraces this openness.

It is only by applying some sort of mapping between the symbols and a pattern generated via a pattern identification process that some sort of meaning can be ascribed to the symbols. Thus a pattern that relates to the commonality of various data that has the property of oneness as discerned by the pattern labelled 'one' that fits the appropriate data can be 'mapped' onto the symbol '1' as used in mathematics.

This mapping enables a pattern to be manipulated to generate new configurations. For example one might count 23 sheep in one field and 17 and another, then by applying mathematical theorem '23+17=40' one can infer that there are a total of 40 sheep in the two fields.

However just because one can effect a mapping in one situation does not mean that the same mapping can be applied in all situations. The application of an abstract system to patterns can only be considered to be a suggestion for how a pattern can be manipulated. It is necessary for the conclusion to be checked back against original data to check that the mapping is accurate. Only when the application of an abstract system to patterns consistently demonstrates accurate results, as in the adding of numbers of sheep, is it reasonable to gain confidence in the mapping and feel assured that it will apply in most cases. In other words, the mapping can be labelled as 'true'. And then it can also be declared as 'true' that '2 sheep + 2 sheep = 4 sheep'.

An example of a situation where a mapping of this form must be treated with caution is in the case of drops of water where 2 drops of water + 2 drops of water does not necessarily = 4 drops of water; but instead 2 drops of water + 2 drops of water = one small puddle.

Mathematics can also be useful for generating templates which can then be used as a starting point for a pattern identification process. An example of this would be in searching for a pattern relating to the starting height of a solid object and the time it takes to reach the Earth. A template in mathematical form that could be used is: h=a(t*t) +bt +c. Then a pattern could be found of the form h=a(t*t), where h is a measure of the height above the ground and t is a measure of the time it takes to reach the Earth and a is a constant whose value would depend on the units chosen for the measurements.

It is only the abstract systems that generate, what one might call, 'interesting' theorems that are of interest. And by 'interesting' I mean ones that either have a practical use or are stimulating in some way and have the potential to be of practical use.

The abstract system of mathematics is clearly of high interest. Conway's game of life, another logical abstract system, is also interesting.

One can also construct an abstract system which incorporates the contents of an English dictionary within its axioms. With a suitably devised set of logical rules which act upon the English dictionary and treating each of the words in the dictionary as a string of abstract symbols, it would be possible that it could generate theorems of the form 'all sheep are mammals' and 'there are no married bachelors' and so on. These theorems could be considered to be 'true' within the system.

These theorems consist solely of strings of symbols and while within that abstract system, they are entirely meaningless. It is only when there is a mapping between those strings of symbols and the labels of patterns in the brain that what could be construed as 'meaning' can be ascribed to those theorems.

Whether such a system could produce theorems that could be considered to be interesting remains a moot point. For most people would already know that there are no married bachelors, as least given the normative meanings of the words. So they would not gain any insight from an abstract system arriving at that conclusion. However outside the normative dictionary meanings, it is possible that 'bachelors' is the name of a musical group consisting entirely of married women. In which case a woman would be married bachelor and the mapping from the theorem 'there are no married bachelors' to the real world would be invalid.

The abstract system as a model for logical systems such as mathematics which do not have any direct connection to the senses is both simple and powerful. It allows for a large degree of freedom in the selection of the axioms and rules, the only constraint being that they are able to generate theorems that are interesting.

This model is also entirely compatible with the pattern identification process previously described. It negates the requirement for mathematics to be inherently and objectively 'true'.

Abstract systems are, by their very nature, logical. They are defined by their axioms and their rules of inference. The theorems that they generate are necessarily logical inferences from those axioms.

The abstract system model has a number of advantages, some of which I will list here:

1. It is a simple model

2. Its foundations are explained

3. Their links to the pattern identification processes and hence to the labels in the pyramid of patterns explains how they fit in to the rest of philosophy.

4. It removes the dichotomy regarding truth of mathematics and other logical systems.

5. It is general enough to incorporate mathematics and other logical abstract systems.

6. It avoids the problems that a philosophy of mathematics using a different paradigms typically encounters such as completeness, infinitism and provability.

An abstract system can be considered to be a logical system. In order to generate theorems it requires some form of logical machine. Typically this would be the brain of a person but it could also be mechanical or electronic;

a computer is a good example of the sort of electronic machine which can be programmed with the axioms and rules of the abstract system and then set running to generate theorems. It could be set running to generate all possible theorems of the system; though it is quite possible that if set to do so, it would never halt as there could be an infinite number of possible theorems. Or it could be set to apply specific applications of the rules to specific inputs in order to output specific theorems.

The theorems of an abstract system can be labelled as 'true' to indicate that they are indeed theorems of that system which would thus distinguish them from strings which are not theorems of the system. Thus for mathematics the string '2*5=10' can be labelled as 'true'.

However it is possible that a different abstract system could use similar symbols to that of mathematics but which have a different significance within that abstract system, in which case it is quite possible that the string '2*5=23' could be a theorem of that system and hence could be labelled as 'true' within that system.

However, in practice, this is usually not the case as the particular symbols used for a particular abstract system are generally unique and hence can identify the particular abstract system to which they belong. Thus the use of the symbols '3', '*' and '=' would normally indicate that they come from the abstract system of mathematics; while 'married' and 'bachelor', if being treated as elements of a logical system, would indicate that that abstract system is one that incorporates the contents of an English dictionary.

One of the tasks of mathematicians is to select and organise the axioms and rules of that abstract system. For while there is no particular requirement for the suggestion of an axiom or rule within an abstract system to conform to a particular style, it is necessary that the suggested axioms and rules are compatible with the ones already in existence. For should the axioms and rules not be compatible with each other, it would be impossible to construct a working machine, even a hypothetical one, which has the capability of generating interesting theorems.

Or to put it another way, one of the tasks of mathematicians is to search for axioms or rules which can be added to the system and which are compatible with the existing axioms and rules and which are likely to enable the generation of interesting theorems. Over the years this is what has happened; mathematicians have added new symbols and rules such as sets, matrices, complex numbers, calculus and so on.

It is interesting to compare the pattern identification processes as applied to sense data with the logical abstract system processes which follow from axioms. The human brain is very adept at finding patterns. Presumably this is because pattern identification processes are hardwired into the brain. In contrast, the human brain struggles with the much simpler algorithms of abstract systems. School children typically struggle to learn the concepts and processes of mathematics. Typically they begin by learning applied mathematics such as '1 sheep + 3 sheep = 4 sheep' rather than the pure mathematics of symbol manipulation.

In the TPP model there are only two systems for drawing inferences. These are the pattern identification logical process and the abstract system logical process; though the two may be combined.

CHAPTER 4

Creating patterns

We have now identified two processes of inference and while they have similarities, both being based on logic and algorithms, they also have a significant difference in that one of them, the real system, uses sense data as its input foundation, the other one, the abstract system, uses axioms as its foundation.

The main system used throughout this book is that of the real system. All conclusions relating to the real world are ultimately derived from sense data, albeit perhaps through many levels of pattern identification. The main use of abstract systems is for the purpose of generating templates that can be used in real systems for finding best patterns. They can be regarded in this manner as a source of imagination.

It should also be noted that it may take more than one pattern to model a set of data effectively. It may be that a primary pattern is identified but when the data is reconstituted from the pattern and compared with the original data there may be what are called 'residuals', the difference between the theoretical (pattern-based) data and the actual data. These residuals can then be used as input data for a secondary pattern identification process to search for a pattern that will fit the residuals. If such a pattern is found it would then be a secondary pattern to the primary one. And should the application of both the primary and secondary patterns lead to further residuals when compared to the actual data, then a tertiary pattern can be searched for and so on.

As an example these levels of patterns can be found in the physics of a solid object falling through the atmosphere. The primary pattern would be one of the force of gravity accelerating the object towards the surface of the Earth. A secondary one would be a force in the opposite direction owing to air resistance. A tertiary one would be the effect of friction of the air molecules brushing against the sides of the object.

Throughout this book and in the construction of the TPP model of the world it is primary patterns that will be identified as they constitute the fundamental structure of the model. And it is envisaged that these primary patterns are in essence universal to all people and hence applicable to all people and their experiences of the world. However it may well require the application of secondary or higher level patterns to model accurately the data of their particular experiences.

The significance and importance of patterns and the process of pattern identification is that they constitute data compression. An immense amount of data from the senses can be compressed into data that is simple and manageable. More than that a pattern can draw out information about the source of the data in a way that no other process can. In other words, a pattern that has been identified from sense data not only constitutes a means of compressing the data but also says something about the nature of the source of the sense data. And this information can be used to create higher-level patterns.

Another useful consequence of creating a pattern is that it can be used to interpolate between data points. So that if there are gaps in the original data, a pattern identified from the data can be used to suggest possibilities for those gaps.

Also patterns can be used to predict possibilities beyond the range of the original data. Its ability to make predictions is not a necessary requisite for the merits of a pattern, but if a pattern can be used to recreate the data beyond the range of the original data and hence make a prediction and if that prediction turns out to be accurate then this would be a confirmation of the accuracy of the pattern.

One of the intriguing things about the process of identifying patterns is that there is no one simple algorithm that can find all possible patterns. Instead what is typically required is a template to act as a seed for the process. But where does the seed come from? One possibility is to take a small sample of the data as a template and determine whether it repeats throughout the data. Or perhaps a template that has worked well for a different pattern can be used as a seed for a different domain of data. Or possibly a random seed is used. Or perhaps two different templates which have been effective previously can be merged to create a third. The determination of a seed to use as a template for the pattern identifying algorithm is akin to what we might call 'imagination'.

Once created, a pattern can be stored and labelled for easy access. The labels for many of the highest level patterns have the unusual property of being able to be spoken, written and otherwise communicated. But this is not the case for lower level patterns such as those used for creating a 3 dimensional model of the world from 2 dimensional visual information nor for the specific instructions for wiggling one's left big toe.

I have discussed in some detail the logical processes of creating a pattern for the reason that these logical processes underpin all our knowledge of the world.

Chapter 5

Interacting with the world

A brain or other data processing device with access to sense data can create patterns from the data and can continue recursively to create ever higher levels of patterns. However this is only part of the story, for a brain also has access to output motors with which it can interact with 'the world'. And this interaction is essential for creating a model of the world. For without this interaction, the patterns that are created from sense data would be no more than meaningless and abstract pictures.

A brain does not interact with the external world solely by obtaining data from its senses; it also has the ability to interact with the external world through control of its output messages to motor devices.

Though of course initially the brain is not aware that that is what it is doing. All it can detect is that if the signal is sent to a particular output, then the data received from its senses is altered in some way. For example, when a signal is sent to the 'eyes' to 'swivel left', the sense-data received from the 'eyes' will alter in a systematic way which is reversed when another signal is sent to the 'eyes' to 'swivel right'. (I have used inverted commas here to reinforce the notion that the brain does not, at this early stage, have any concept of what eyes actually are, nor of the concept of left and right.)

What the brain then has to do is make an association between the patterns in the data received by the senses and the patterns of the signals sent to the motor devices.

Also, signals can be sent to other outputs that control, for example, limbs which can affect its 'motion' which will again influence the sense data in a systematic way. 'Objects' can become 'larger'; 'sounds' can become 'louder'; 'touch' can become 'coarser'.

It is this cross-correlation between the sense-data received and the motor outputs that gives 'meaning' to the patterns and allows for the beginning of the long process to make sense of the world.

Two of the most basic patterns that are acquired and are essential in making sense of the data are those of 'time' and 'space'.

The brain can then go on to combine data from its senses. Patterns formed from seeing, hearing, smelling, touching and tasting can be combined to build up a composite picture of the world.

The brain can also continue to generate successive layers of higher order patterns which are created using the lower order patterns as input data.

And as the pyramid of patterns grows higher, at some stage the brain might learn to distinguish between things which 'exist' such as trees and rocks and those which do not exist such as dreams and optical illusions. Those things that are considered to 'exist' have a commonality which distinguishes them from things that are not considered to 'exist'.

And at some even higher level in the pyramid of patterns the brain may consider that the pattern represented by 'I am' fits a particular section of the pyramid of patterns and it would become aware of its own existence. It is perhaps at this stage that one might refer to the part of the brain that is self-aware as a 'mind'.

At another stage a brain may create patterns for the cycle of birth, life and death for all life forms. Then the brain might decide to look for patterns for the birth of life itself and start searching for data that might relate to this. Having amassed sufficient data relating to the beginnings of life and its progression to the present day it could use this as input data for a pattern identification process; it is likely that it would eventually identify a process of evolution as being the best, if not the only, pattern that fits the data.

In essence, evolution is a very simple and elegant theory and yet extremely powerful in its ability to explain the data for so many facets of life. Evolution constitutes one of the anchor points of TPP. I shall go over some of its main points here to avoid any misunderstanding.

In essence genetic variation leads to variation in the morphology of an organism. Some organisms breed successfully, some do not. That is all there is to it (at least at this level of discussion). There is no particular 'selection process' and no identifiable quality of 'fitness'.

Though this process of evolution is extremely simple, it has resulted in the many variations and complexities of life on Earth. It explains why everyone is different from everyone else and also why there are so many similarities. It would seem reasonable to presume that most, but not necessarily all, of the phenotypes that an organism has are somehow useful for the replication of its genetic code.

And the primary use of a brain is undoubtedly for the purpose of making decisions and to translate those decisions into useful actions so that it may have a better chance of replicating its genes.

The brain is a complex organ and only part of it is dedicated to the creation of a pyramid of patterns. Some of the decisions that the brain makes may be mechanistic or instinctive in nature and below the level of the brains self-awareness. These will include decisions regarding heartrate and temperature control.

But whatever the form that the decisions made by a brain takes, it would seem that it has no other function. Hence the assertion that the brain is a decision making organ would seem to be reasonable and hence it constitutes an anchor point for the pattern paradigm.

Mechanistic linkages in a brain can work in the same way as a thermostat or a computer in the way that it follows the logic of the form 'if A then X'. This can be seen in the form of the body's temperature control which can be modelled by something like 'if the temperature gets too high then sweat, if the temperature gets too low then shiver.' No model of the world is required, it is a simple mechanistic relationship.

However if the process of data input to action output requires the inclusion of a model of the world then this is not so simple. For then the logical process must pass through the pyramid of patterns, for only in the pyramid of patterns can there be a model of the world.

If the input data is, for example, a sensation of hunger, it is possible that for a primitive organism without a model of the world or pyramid of patterns, a purely mechanistic response could be initiated, which could be of the form 'move forward and chomp the jaw muscles.' And one can imagine that in a very primitive environment this strategy could prove effective. However it is not efficient and using a model of the world to determine the direction of motion and the timing of the chomping of the jaws would certainly be beneficial. Thus an organism with a pyramid of patterns and a model of the world can actively search for food, identify objects as possibly being food and then eat it, which is far more efficient.

There can be no direct logical linkage from the sensation of hunger through the pyramid of patterns to the consumption of food, for a pyramid of patterns is too complex to allow for a direct linkage. So then the question arises how is it possible for the organism to respond to the sensation of hunger by finding and eating food? The answer must be that it has a model for its sensation of hunger and the world outside and the knowledge that the sensation of hunger can be satisfied by finding and eating food.

And there are other physical signals that the organism needs to respond to in order to maintain its physical well-being. These might include being cold in which case the organism can seek shelter; or it might feel in danger in which case it might seek out the company of close relatives.

But of course the sensations of hunger or cold are not merely pieces of information to be acted upon or not depending on a whim, there is associated with them a sensation of pain which demands action and then upon the easing of that pain there is a sensation of pleasure such as in eating food or gaining warmth.

So then the part of the brain associated with the pyramid of patterns seeks to find pleasure and avoid pain.

This then provides the third part of the logical process of an organism extending past the mechanical 'if A then X' to a more complex process that incorporates a pyramid of patterns and a model of the world.

The first part, as discussed previously, is the creation of a pyramid of patterns from sense data, the second part is interacting with the world using that model of the world and the third part is the motivation to interact with the world through sensations of pleasure and pain.

So what the mind (the part of the brain that incorporates the pyramid of patterns) is motivated to do is to achieve pleasure and avoid pain. It does this both for the short-term and long-term. For there is pleasure associated with a feeling of contentment in knowing that one has a store of food available for times when food might be scarce. I shall summarise this goal of pleasure and avoidance of pain in both the short term and long term as that of seeking happiness.

As already mentioned people have similarities and also differences, as determined by their specific genetic makeup. So people will undoubtedly have different degrees of pain and pleasure for all the different centres for pleasure and pain. And this will affect the strategies that they use to maximise happiness and minimise pain.

It is of interest to explore the logic of a brain's decision-making, given the three parts of the process: 1: the model of the world, 2: the pursuit of happiness and 3: possible actions. In essence it is one of using the image of the world to test possible actions and then determine which one seems likely to bring the most happiness and least pain, both in the short term and in the long-term.

Decisions would have to be made regarding whether it is better to endure a little pain now for the procurement of much happiness in the future or not to take the risk and opt for the immediate pleasure and ignore the long-term consequences. This is a choice for the particular person to decide. And it will depend in part on their confidence in the accuracy and comprehensiveness of their model of the world. For if they believe that their model of the world is accurate and comprehensive then they may be confident that that future pleasure can be attained, whereas if they have little confidence in their model of the world then they may consider that the prospect of future happiness probably won't occur anyway and in which case they may consider it a better option to choose the immediate pleasure.

An algorithm for this decision-making process might look something like this:

- Consider all possible courses of action.
    - Take each action one at a time.
        - Use one's model of the world to predict possible or likely outcomes of the action in both the short-term and long-term.
        - Evaluate the pleasure and pain that is likely to be associated with each part of these likely outcomes.
        - Arrive at the total value of net pleasure and pain for this course of action.
    - Repeat for each possible course of action.
- Select the action that has the highest net value pleasure less pain.

This is a somewhat idealised algorithm as it does not incorporate any consideration for the time required to complete this process. For often decisions are required to be made on a real-time basis and where it is important to make a decision quickly. And the processing of an algorithm of this form can be quite time consuming, particularly in the consideration of the possible long-term consequences of a particular action.

If time is a factor then this process will have to be simplified or perhaps an action that was reasonably successful in a previously encountered situation which is similar may be chosen.

The decision that is finally chosen is up to the individual person who makes it and they have no recourse except to live with the consequences.

There are no guarantees for the consequences neither are there any guarantees that the projected happiness will eventually be enjoyed.

It is this concept of the motivation of a mind to achieve happiness that constitutes another anchor point for the TPP philosophy.

I should note here that I'm using the word 'happiness' as a label for 'that to which the mind aspires'. And there are certainly sufficient similarities with the normative meaning of the word 'happiness' to justify its use and to preclude using a different word.

I should also note that when a person makes a decision that they consider will maximise their happiness, it is not what will necessarily actually bring them happiness nor is it what other people think will bring happiness but rather is it is their particular perception of happiness which is used in their decision-making process.

Chapter 6

Communication

So far what has been discussed would be applicable to many non-human animals. But what follows only really applies to humans as it is only humans who are known to have developed an advanced language for communication.

The base elements of a language are words and words are labels for patterns or sometimes a group of patterns.

Since different people create different patterns in their pyramid of patterns, the same label will indicate slightly different patterns for different people.

A trivial but obvious example of this is for colours. People generally do not agree on the range of colours that can be labelled 'red'; and that is just a simple word or pattern. (One person's red could be another person's pink.) Certainly for more complex patterns such as justice, freedom and honour there will be even greater differences between the patterns that one person has put under one label and the patterns that another person has attached to the same label.

Yet communication is possible. This ability to communicate, first with spoken language and then with written language has had a major influence on the human race. It has enabled the exchange of ideas and information which has led to a great leap in technological development, particularly since the advent of writing.

Then the question can be asked how does this communication influence what we have already discussed about pattern identification and decision-making?

An incoming communication can be treated like any other sensory input; the sound or writing is identified in the pattern of the overall communication and is processed through the pyramid of patterns so that it can be evaluated and a decision made as to whether to store it all, learn from it or perhaps act on it immediately. In this way it is similar to any other sensory input.

But it is also different from ordinary sensory input as it is a communication from a person and the pattern it invokes may not necessarily be classed with other patterns that have a direct link to the 'real' world of sense data.

An outgoing communication can be treated like any other decision and action. It is a decision to communicate and is enacted through the jaw muscles and vocal chords for spoken communication or through the hand (or other motor output) for written or typed communication.

And as with all other decisions made by a person the ultimate purpose of the communication is to maximise happiness.

What I want to do now is to discuss some of the logical possibilities of how language and communication can interface with the pyramid of patterns, happiness and decision-making and to show how these tally with the facts of the world.

Since communication works with labels of patterns, once the label, or word, in the communication has been identified it can be linked directly to the associated pattern.

Suppose a person receives a communication that has the phrase at the end "and that is the truth", which would indicate that the speaker attests that the communication is an honest one and represents a portion of their pyramid of patterns. How is the receiver to process this information?

Clearly this constitutes something different from the knowledge that they have garnered purely from their own personal experience and sense data.

The first part of the process is to determine whether the new information, which at this stage can only be considered a possibility or suggestion, fits in with one's pre-existing pyramid of patterns or not.

If it does fit and the speaker is trusted to be honest and have integrity then it could be reasonable to incorporate the suggestion as a belief alongside one's other beliefs.

If it does not fit harmoniously into one's pyramid of patterns, then the outcome is not so straightforward; other factors come into play. It will not be a simple process of finding the best pattern to fit the data.

The problem could be that one's pyramid of patterns does not hold sufficient data that is relevant to the communication. Or perhaps the relevant data that one does have does not support the patterns suggested by the communication.

In either case one might be tempted to simply reject the suggested knowledge. But in deciding whether to do so or not, other factors come into play.

For example, consider the situation of a child. She has little option but to accept the culture and knowledge presented to her by her parents and other teachers. For to do otherwise would court great unhappiness. So she will try to make the best sense of the culture and knowledge of her community even though she has very little relevant personal data to corroborate it. And she might quite possibly search for other data that will support the knowledge that has been presented to her.

This illustrates a number of important principles:

1. The pursuit of happiness supersedes all others. If it is deemed beneficial to adopt a particular belief, irrespective of whether it is the best pattern to fit the data or not, then that belief will be adopted.

2. Adopting a particular belief can be classified as a decision even though there is no immediate motor output associated with it. In all likelihood the adopted belief will have an influence on a decision and motor output at a later time.

3. Altering beliefs is no easy matter. For in a way the pyramid of patterns defines who we are as a person and to invoke radical change is akin to changing one's identity. And if the changes are undertaken, it would take a lot of brain processing time. So to go down that road would require motivation from either pain in the present or the perceived probability of happiness in the future or possibly both.

   An easier course would be to simply adopt the belief and then search for data to back it up and ignore any data that does not fit with it.

It is not my intention here to suggest that any one course of action is better than another, I only wish to point out that the individual, within their own specific environment, has choices and can decide which is the better course of action for them. My purpose here is to highlight some of the logical processes and logical constraints that are applicable to the process of decision-making.

While language and education had been a great boon to mankind both in society and technologically, it has come at a cost; that cost being the potential for a schism between what one believes because it constitutes the best pattern to fit the data and what one believes because one considers it expedient to do so.

Ideally of course, one would not adopt a communicated piece of purported knowledge until such time as one had collected sufficient relevant data and have deduced that bit of knowledge for oneself by identifying it as constituting the best pattern to fit the data. However in practice, this is virtually impossible. The amount of time and effort required is just not pragmatically possible.

The other ideal possibility is that all the purported knowledge that has been communicated to one is actually accurate and would be the best pattern to fit the data if only one had sufficient relevant data and time to effect the process. However this ideal possibility cannot be relied upon to be the case. People have different models of the world and different experiences so even with honest communication the purported knowledge may not fit with one's own model of the world.

Whether one chooses to adopt some communicated knowledge will also depend upon one's view of the person who is the instigator of the communication. Does one consider the other person to be knowledgeable on that particular topic, are they someone one respects and wishes to emulate? If so then this might suggest that adopting such knowledge as one's own may well bring happiness and hence be something that one would choose to do.

Alternatively if the communicator does not command one's respect as a source of knowledge or is not a person with sufficient perceived status for emulation then the purported knowledge is more likely to be rejected or at best be stored and kept in limbo until further data is received and a pattern identification process can be effected.

Purported knowledge that has been communicated can also be tested for veracity by examining particular new data points that one has discovered for oneself and determining whether they are a fit to the pattern suggested by the purported knowledge. If they do, then this is good evidence that the proposed pattern is a good one and can be accepted even though one has not personally determined for oneself that it is in fact the best pattern to fit the data. For example, one might be told by a science teacher that "force equals mass times acceleration". Then one might do experiments using springs and little wooden vehicles and stopwatches to gather some relevant data. And then one might find that this new data does in fact fit with the pattern suggested by the teacher. In this way one has not discerned for oneself that that particular pattern is the best one to fit the data compared to any other proposed pattern, however one has established that that pattern does indeed describe the new data that one has established through experimentation and observation.

Communicated knowledge can also be accepted if its inferences lead to consequences that are undeniable. For example the existence of modern electronic technology is undeniable and so the scientific foundation upon which the technology is based can be accepted even though one might have no understanding of it oneself.

I do not want to delve too deeply into the complexities of communication as it depends so much on the specific circumstances of the people involved. All I want to do is to explore the basic logic of communication and to identify some of the motivations behind it and the constraints which apply. I also want to show that the only major anchor points required to create an efficient model of the process are those of the pyramid of patterns and the pursuit of happiness.

There are two sides to the use of language for communication. There is the instigator or speaker and there is the receiver or listener.

The action of speaking or writing or gesturing is just another use of the motor outputs controlled by the brain and is motivated by the brain's desire for happiness.

So one can use language to give 'instructions' to other people or to educate them all to swap stories or perhaps to spread lies. I have put 'instructions' in parentheses to emphasise the fact that what the speaker considers to be an instruction may not be received by the listener as an instruction. I have also mentioned lies at this juncture to emphasise the point that communication is not necessarily honest. The aim of speaking is to bring happiness (or the avoidance of pain) to the speaker and if it is considered that telling a lie is the optimum course of action then this is what will be effected.

Also the message sent may not be the same as the message received as some messages need 'translating' before they make sense. Also sometimes the literal meaning of the words sent will differ from the intended communication.

For example, someone might say: "All sheep are white". This statement is clearly false, but nevertheless is most likely an honest communication; so one could translate it to: "This person believes that all sheep are white".

Or someone might say: "I will win the lottery this week", clearly this is a speculative statement and does not literally represent the mind-set of the person; so this could be translated to "This person is hoping to win the lottery this week".

I'm only using the broadest brush strokes here to describe how communication works and how it fits in with other concepts of the TPP model.

Communication between people using a comprehensive language is perhaps the most significant difference between humanity and the rest of the animal kingdom. It is responsible for the huge advances in technology that have radically altered the daily lives of people. It has achieved this by, in effect, allowing one's person's pyramid of patterns to have access to those of another person.

Finding the best patterns from basic sense data is a time consuming process and communication allows for specialisation so that one person can focus on finding the best pattern for a specific and limited range of data and then communicate their findings in the form of patterns and the domain of those patterns to another person. Then assuming that the pattern indicated is efficient and potentially useful, the listener's knowledge of the world will increase with little effort on his behalf.

These communications can be effective not only between people in direct contact but with the introduction of writing they can also be effective through indirect communication over the ages.

However, as with all communications, their accuracy may be tempered by the fact that people communicate with the intention that they will bring some degree of happiness to themselves; with this in mind the possibility that some communication may deliberately be inaccurate or biased must also be considered.

CHAPTER 7

Decision Making

In this chapter I want to combine and extrapolate from the anchor points previously identified and show how they can be used to describe the world.

Decision-making is at the heart of the experience of life and an understanding of it is at the heart of philosophy. It is the brain that is the body's decision-making device and which is the interface between sense data inputs and motor outputs.

It was previously discussed what the logic of decision-making would entail and how that logic would apply to all animals. I now want to discuss how language and the degree of communication that is unique to humans impacts on the decision-making processes and some of its consequences.

Decision-making is a complex process and in the complex world of nature and social interactions it can be extremely difficult to make decisions that look beyond the immediate gratification of pleasure.

In an ideal situation, a decision-making process would consider all options and all possible outcomes and how those outcomes might affect other people and how they might respond to it.

Decision-making is at the heart of animal life and at the heart of what it means to be alive.

However the decision-making process is a difficult one despite being fairly straightforward. For decision-making can require a large amount of computing time; the time it takes a brain to complete the process. And an adequate amount of time may not always be available, because people live within a real-time environment. The situation and immediate environment are continually changing and decisions need to be made within this real-time environment. So all too often the decision-making process has to be curtailed or simplified because of time constraints.

But first let us consider the decision-making process in an idealised environment where there are no time constraints.

In this idealised environment an algorithm which could act as a model for the decision-making process would look something like this:

1. Gather all relevant information.

2. Make a list of all possible decisions or actions.

3. Take each possible decision one at a time.

4. Use one's model of the world to predict the short-term and long-term consequences of the possible action associated with that decision. Included in this assessment would be the impact of the action upon other people and how they might respond and how those responses might impact upon oneself. For each of these possible consequences estimate a probability that that consequence would eventuate.

5. For each of those possible outcomes ascribe an estimate for the pleasure or pain that is likely to impact upon oneself as a consequence of that outcome and of enacting that action.

6. Apply a total happiness quotient for that action which would be a net total of the pleasure and pain associated with each

outcome multiplied by the estimated probability that that outcome will eventuate.

7. Repeat the steps 3 to 6 for each possible action, so that one ends up with a net happiness for oneself that is associated with each possible action.

8. Select the action that has the highest net happiness associated with it. It is possible that this could be a negative value should one be in a dire situation, in this case one would choose the least negative one.

Note that this crude algorithm follows from the previously identified anchor points of the brain as a decision-making device and that the brain seeks to maximise its own happiness.

Also note that this process is dependent on the person's perception of the world and one's perception of the consequences and how other people will be affected by the consequences. It is not dependent upon what other people think will be the likely outcomes of an action. Nor is it dependent upon what other people think will bring them the most happiness; it is only dependent upon what the decision maker themselves consider will bring them the most happiness. In this respect the decision-making process is highly individualistic and subject to the particular vagaries of the person's pyramid of patterns and their particular character as well as their situation.

The accuracy of the evaluation of the possible consequences of an action will depend a great deal upon the accuracy and extent of the decision-makers model of the world, and these in turn will depend upon their learning about the world and their experience in applying that learning.

The evaluation of the long-term consequences of an action will of course become increasingly uncertain with the increasing depth of evaluations. And the degree of the depth of evaluation will depend upon the decision-maker's confidence in the accuracy of their model of the world.

In practice however, a deeper level of the analysis of the expected consequences of the decision and its associated action is most often not possible. For there may be considerable time constraints and also many uncertainties in predicting the possible outcomes of an action.

Also there is an element of trial and error, if a decision leads to beneficial results then should a similar situation arise then a similar decision would seem sensible. Whereas if the earlier decision led to unfavourable results then should a similar situation arise a different option might be selected.

It is often the case that it is the long-term consequences of decisions that can have the greatest impact upon a person's happiness. Yet if a person's analysis of the consequences of their decisions never reaches the depth of the long-term consequences, then they will only focus on the short-term consequences of an action and its associated happiness to the detriment of their possible long-term happiness.

Or at least this would be the case if people made decisions based entirely on their own personal experience and their own personal knowledge of the world. But through communication people can learn from others and their experiences and listening to their advice could prove beneficial. For this can provide a shortcut to the evaluation of long-term consequences which allows for consideration of the long-term consequences and also speeds up the decision-making process.

This might take the form of people relating their experiences in love or war or exploration, or it could take the form of an ethical code of recommended behaviour. Or it could be simple advice from a friend or parent or other respected personage. These communications of experiences, advice and morals can act as a guide to modes of behaviour and the evaluation of long-term consequences of decisions that might otherwise be hard to determine from one's own experience and one's own processing of the relevant data. However these communications have no merit or significance beyond their worth as a communication. The value of the communicated advice or morals can only be determined by the receiver of such communications; and this is a decision to be made by the receiver.

For choosing to believe something constitutes a decision. It is a decision about whether or not to add that belief to one's pyramid of patterns.

When generating patterns from raw sense data there is only one way to do so and that is finding the most accurate, simple and efficient pattern that fits the data. Consequently the pattern automatically becomes a belief, or at least a working hypothesis. But with the advent of language and the ability for communicated ideas to be slotted directly into one's pyramid of patterns at quite a high level then a different criteria with which to test the validity of that pattern as being the best pattern possible is required.

Some of those criteria might include how well the idea (or pattern) fits in with other previously accepted ideas (or patterns). Also there may be corroborating evidence such as useful technology which clearly emanates from the theory (such as the use of a cell phone which would indicate the validity of theories of electronics) or there may be specific experiments one could conduct to identify a particular data point which is predicted by the theory.

Or it may be that the suggested theory was proposed by people one respects such as teachers and parents.

But ultimately the criteria for the adoption of a suggested theory or pattern is: will the adoption of this theory as a belief lead to greater or less happiness?

While a process of this form may be simple enough, the actual logic beneath it remains unknown. For it neither qualifies as a pattern identifying process nor does it qualify as an abstract system based upon known actions and methods of inference. And because it is opaque it does not qualify as an objective process of inference at all. Presumably it does follow its own internal logic however this logic remains unknown and is also specific to the particular person making the decision.

The decision-making process can also result in the generation of a communication in the form of a statement. All statements can be considered to be "opinions", because the processes that produce the statement are based upon actions and beliefs and intentions that are hidden.

However, a statement, despite being an opinion can refer to ideas, patterns and logical inferences which are not obscure and hence the idea behind the statement can be evaluated in the brain or mind of the receiver. This would be done through one of two processes of inference: that of pattern identification or that of an abstract system.

So on hearing the statement "the sun is shining", the receiver of the communication can confirm this, or not, from their own sense data and patterns created from them.

If the communicated statement was "the square root of 169 is 13", then this opinion can be confirmed, or otherwise, by exploring the inferred system of mathematics and seeing whether this statement is compatible with a theorem of that system. While statements about mathematics are opinions; the creation of an abstract system such as mathematics with its axioms, processes of inference and theorems constitutes an objective system since inferential processes can be reproduced by another person.

And so also with most of science. Statements about science can be considered to be opinions, yet the ideas about science can be reproduced by students of science doing experiments which can confirm, or otherwise, the patterns suggested by statements about science.

So also with philosophy. Statements about philosophy are opinions. However there is a difficulty with much of philosophy in that very few of the ideas can be confirmed by doing experiments, or by reproducing them using an abstract system or by creating patterns directly from sense data. Some philosophical paradigms have tried to use logic and language as a process of inference however this constitutes a conflation of two different processes and can generally be regarded as unproductive. This will be discussed further in the next chapter.

Another point to note about the decision-making process is that not only are the details of a particular process opaque to other people, they are also, to a lesser or greater degree, opaque to the decision-maker themselves.

For there is no requirement that a record be kept of all the details of the decision-making process or even that it would be possible for the details to be recorded. In the same way a pocket calculator is not required to describe, and is indeed incapable of describing, the processes by which it arrives at the answer to a particular configuration of input keys such as "2", "+", "5", and "=".

Any 'justification' a person gives to making a particular decision, even if honestly given, may not be entirely accurate and certainly cannot be relied upon any more than any other opinion.

The term 'decision-making' refers to every decision made by the conscious mind that can have an influence on the motor outputs under its control. Thus decisions can cover the range from scratching an ear to buying a house and beyond. It includes the making of statements and even the choosing of beliefs.

However the decision-making process is one that is typically based on limited knowledge, not only of the possible options but also of the possible outcomes and in particular the limited knowledge of how the possible outcomes would lead to personal pleasure or pain. There may also be a time constraint which could limit the exploration of some options and their consequences.

So although the decision-making process may well follow a logical path it is one that must explore many possibilities before selecting one particular decision. In this way the process can be more effectively modelled using game theory rather than a strictly linear logical one.

It seems likely that mankind is unique among species in having a fairly good idea of how long a person may be expected to live, under normal circumstances. This knowledge is useful in setting a limit to the exploration of long-term consequences of possible decisions.

How deeply one explores the long-term consequences will also depend upon one's confidence in one's model of the world, for there would be little merit in exploring the long-term consequences if one does not have confidence that these will come to pass. In such a case consideration would likely focus on the short-term consequences and their associated pleasure and pain.

However if one has great confidence in one's model of the world and confidence in one's ability to predict the likely consequences of possible decisions, then it might be more beneficial to take them into account and plan long-term benefits and long-term avoidance of pain.

Decision-making is at the heart of our lives. It defines who we are and what our lives will be. It needs to be at the heart of any meaningful philosophy.

CHAPTER 8

Logical Systems and Logical Processors

Many people have the expectation that philosophy deals with truth, and some readers may be wondering how truth can be extracted from the pattern paradigm hitherto described. However it is not the role of philosophy to be of itself true so much as it is its role to explain the concept and origin of truth.

Some form of logical system is essential for all inferences and decisions. And at the heart of all logical systems there is a simple logic based on axioms and rules of inference.

And for every logical system there is a logical processor which produces the inferences, the theorems or the decisions.

At its most fundamental level, logic is the manipulation of symbols according to particular specified rules. The system is defined by its axioms and rules of inference. The only restriction on the axioms and rules of the system is that they must be sufficiently clear, consistent and comprehensive so that a logical processor can be constructed which follows precisely the axioms and rules of inference and which is capable of generating theorems.

A logical processor needs some instructions before it can generate theorems. In the same way, a pocket calculator needs some of its buttons pressed before it can produce a result. The instructions that control a logical processor can be considered to be an algorithm, for an algorithm is just a set of instructions for a logical processor. While it is the rules of an abstract system that determine the ways that the symbols of the abstract system can be manipulated, an algorithm determines how those rules are to be applied and in what order. A mathematical proof is an example of an algorithm which describes a set of rules to be applied in order to generate a desired theorem.

A system with its symbols and rules of inference is inherently meaningless; in isolation the symbols are no more than symbols. Typically such a logic system has no connection with the real world, it is an abstract logical system.

However, some logical processors can be connected to the real world through some form of sensory input and motor output. A brain is an obvious example of this and a modern robot is another. However, each of these is representative of a very different logical system. A brain can create a model of the world through a pattern identification logical system based on its sense data. In contrast a robot responds to its stimuli in an entirely mechanical way, it does not possess a model of the exterior world.

Logical processors can also be divided into two distinct categories. The first is one whose underlying logic is known, the processor is overt. Typically these are computers (or robots) with known software and hardware which have been designed according to some known logical system. The other type of logical processor is one whose underlying logic is not known, the system is opaque. It is one that has come into being through a process of evolution, not design. But this does not mean that its internal logic is any the less logical for that. The brain is a typical example of this. The only way to find out what the internal logic of such a processor is, is to gather data about its inputs and outputs and then try to infer what the logic must be. This method was used in The Pattern Paradigm to infer that the logical process for perception in the brain is one of pattern identification.

While it may appear obvious that the brain can be modelled as a logical processing device, it does not follow from the previously defined anchor points, hence I shall identify it at this point as being an additional anchor point for the pattern paradigm.

A pattern identification processor seeks to find the best pattern that fits the data. It has as its inputs quantised data (most likely from the senses) and requires an initial template or theory which can be used to try to fit the data. It follows a logical process which can be modelled using algorithms.

Physics is an excellent example of this process where hypotheses are tested to see if they constitute the best fit to the data. The best patterns are the simplest ones and ones that can (perhaps together with other previously identified patterns) most accurately and efficiently reproduce the data. If a pattern cannot reproduce the data reasonably accurately then it cannot be considered to be a good pattern or theory.

If a theory can reproduce the data then it can also be used to interpolate between data points and also to extend the pattern beyond the initial domain of the data. The extension of the pattern could be into future time or some other dimension or quality.

The patterns in physics thus created, such as for time, electrons or force are all best fit patterns; but there is no guarantee that some better pattern might not be identified at a later time.

The logical processes involved in finding the best patterns in science and particularly physics are overt, they are open and identifiable. It is possible for anyone to follow the logical steps and arrive at the same conclusions (assuming they have the same or similar data available).

It is only through pattern identification systems that a direct connection to the real world can be achieved. In contrast, abstract logical systems are entirely abstract; they have no direct connection to the real world. In order to make a connection, there needs to be a mapping between the elements of a theorem and the elements of a pattern from a pattern identification system.

Thus the symbol '1' of mathematics can be mapped onto a pattern that has been labelled 'one' in a pattern identification system.

And typically it is these mappings which allow for the determination of whether the abstract system has generated theorems that are considered to be interesting.

In particular an abstract theorem can be mapped onto a template which can then be used to search for a better pattern for a particular set of data. In this way abstract systems can be extremely effective in suggesting possible templates and this can be seen as a form of imagination; generating possibilities for the real world from an abstract world.

The abstract system of mathematics has proven to be extremely effective in generating templates for physics for example in the fields of calculus and complex numbers. It is extremely unlikely that such complex templates could have been found without the abstract system of mathematics.

Abstract systems, such as mathematics, cannot exist solely as a set of axioms and rules of inference, they also need a logical processor that is capable of generating theorems according to the axioms and rules. This requirement puts constraints on the axioms and the rules of a particular abstract system, for they must be sufficiently compatible with each other that a logical machine can be constructed which incorporates all the axioms and all the rules and which is also capable of generating theorems.

Mathematics is a remarkable example of an abstract system in that it has many varied axioms and rules all of which are compatible with each other and it also generates theorems which are interesting and which are compatible with each other in their interpretation and mapping.

An abstract system is a manipulation of symbols according to the rules of the system. On their own these symbols are entirely meaningless. Their only significance is in the way that different symbols are treated differently within the abstract system according to the rules of the system.

And while mathematics is the most obvious example of an abstract system, there are many others. And in fact anyone can create an abstract system. All that is required is to create some axioms and rules of inference. But this does not mean, of course, that this new system would be capable of creating interesting theorems.

To illustrate how the symbols are inherently meaningless one could create a variation of the mathematical abstract system in which theorems are generated of the form '2+2=4' and '2+2=5'. This does not necessarily indicate that there is an inherent incompatibility or contradiction within the system itself. All it would mean is that the theorems of such a system would not map neatly onto, for example, sheep in a field. In other words the mapping of the symbol '4' to the pattern labelled by 'four' and the mapping of the symbol '5' to the pattern labelled by 'five' is unlikely to produce interesting conclusions.

The concept of truth when associated with abstract systems is somewhat different from the concept of truth as applied to patterns. Truth as regards to abstract systems can only be used as an indication of self-consistency within that particular abstract system.

So the axioms of an abstract system can be labelled as 'true' albeit only within that system. So also the theorems of a system can be labelled as 'true' to indicate that those theorems have been generated by that particular system. This is useful as the label can be used to distinguish strings of symbols which are theorems of the system from those which are considered not to be theorems of the system.

Thus within the system of mathematics, the string of symbols '2+2=4' can be labelled as 'true' as it can be considered to be a theorem (or possibly an axiom) of the system. Whereas the string '2+2=5' is not a theorem or axiom of that system and hence can be labelled as 'false' within the system of mathematics.

However typically the use of specific symbols tends to indicate the particular abstract system to which it refers. Thus the symbols '2', '+', '=' are typically only found within the system of mathematics and so any string containing those symbols may be considered to be referring to the system of mathematics. And thus it is generally not required to claim that the string of symbols: '2+2=4' is true only within the system of mathematics because the system of mathematics is already implied by the use of those symbols and so people may well claim that: '2=2=4' is true without the addition of the rider 'within the system of mathematics'.

The prime use of abstract systems is to produce theorems that can be used to suggest possible templates for use in a pattern identification system. For there are no algorithms that can generate possible templates beyond randomness and variation of pre-existing templates. So if an abstract system can generate theorems that can be used to generate possible templates then this is an important use of an abstract system.

It is important to realise that the mapping between the elements of an abstract system and the labels of patterns is a somewhat haphazard one. It is heuristic; a matter of trial and error.

While it may seem obvious that ordinary numbers such as 1, 2 and 3 are useful for counting sheep and that calculus is useful for calculating the orbits of the planets there is no intrinsic reason why these mappings work and that, say, the inverse does not. So it is a matter of trial and error to see which mappings are useful and which are not.

An example of how a mapping between mathematics and the real world of patterns is inappropriate is that of Zeno's paradox of Achilles and the tortoise. For Zeno's suggestion for the use of mathematics in that scenario does not lead to an increased understanding of the situation but instead leads to an apparent contradiction or paradox.

An interesting example of a hypothetical abstract system is one that includes among its axioms the entire contents of an English dictionary together with all the rules of grammar and syntax. Such a system could then be loaded into a logical processor together with some algorithms for outputting theorems which combine various words in the dictionary. In this scenario, all words in the dictionary are treated as nothing more than strings of symbols. While some strings of symbols might trigger particular switches within the logical processor such as 'is' or 'not' the strings have no other significance. They have no mapping to labels of patterns and are entirely devoid of any meaning.

Nevertheless it is quite conceivable that such a logical processor could output as a theorem the string 'there are no married bachelors' based purely on the relations between the words as specified in the English dictionary which were incorporated into the axioms. This string could then be labelled as 'true' within this particular abstract system. In popular philosophy it has often been claimed that this string is a priori true or that it is analytically true following an analysis of the meanings of the words. But the process by which the conclusion is reached is not logically explicit. According to the pattern paradigm perspective, it can be labelled as 'true' because it is a theorem of an abstract system and it is only true within that system. It is a string of symbols which at this stage has no meaning.

Subsequently the 'words' contained in the string can be mapped onto the words that act as labels for patterns and thus the string of symbols acquires meaning. Then the string can be used as a template to look for a best pattern of the relevant data. When this is done, there would be general agreement that this string did denote an effective pattern under most circumstances. In general, 'there are no married bachelors' would be considered to be representative of a section of a person's pyramid of patterns. However there could be exceptions. For example they could be a musical group called 'the bachelors' which consisted of two married couples. In which case the statement 'all the bachelors are married' would be considered to be an accurate representation of this particular situation and hence the statement could be labelled as 'true'. And also therefore the statement: 'there are no married bachelors' would have to be labelled as 'false', albeit only with reference to the pyramid of patterns and not to the dictionary based abstract system discussed above.

This example highlights the distinction between the theorems of an abstract system and the words and statements that are representative of a pyramid of patterns.

As previously mentioned, abstract systems need a logical processor which can take the axioms, symbols and rules of the system and generate theorems. For without a processor to generate theorems the abstract systems are no more than null systems. The logical processor could take the form of an actual mechanical device or an electronic one, such as a computer or it could be a brain.

However a brain's natural function is to search for patterns; strictly logical processing following rules, as in a typical abstract system, does not come so naturally. If a brain is processing an abstract system which includes strings of symbols from a dictionary such as 'c', 'a' and 't', the brain is likely to associate the string with the pattern it has labelled 'cat'. In this way it is perhaps unavoidable for a brain to conflate some symbols of an abstract system with the patterns in its own pyramid of patterns.

However the two systems of logic are quite distinct. To make a connection between an abstract system and a real system of patterns it is necessary to invoke a mapping process.

This mapping process needs to be identified as the mapping process and not a direct connection, for if it is considered to be a direct connection then false conclusions can be reached.

The patterns created by a brain, using a pattern-identifying logical process, cannot be directly transferred to a simple abstract logical system because typically the form of the pattern is unknown. And even if they are known, as in a scientific theory, they still cannot be loaded directly into a simple abstract system since a simple abstract system would not have the capability of processing them as patterns. So when a label for a pattern is loaded into an abstract system it is only the letters used for the label that are transferred, its connection to the pattern to which it refers is lost, for an abstract system can only operate on abstract symbols.

Thus a logical syllogism can only operate on the labels for patterns and not on the patterns themselves.

All too often the problem is hidden as in the syllogism:

P1: All men are mortal

P2: Socrates is a man

Therefore C: Socrates is mortal.

For this example we shall presume that P1 and P2 are acceptable propositions as they are representative of a typical pyramid of patterns.

So in this instance the propositions P1 and P2 are fed into the abstract logical processor as strings of symbols and out pops the conclusion C. The conclusion can then be fed back into the brain's pyramid of patterns and typically it is found that this is fine and it fits in with the pyramid of patterns and that the abstract logical system, despite operating only on the words as strings of symbols has produced a sensible conclusion.

However, this is not always the case. Consider the syllogism:

P1: A peanut butter sandwich is better than nothing

P2: Nothing is better than perfection.

Therefore C: A peanut butter sandwich is better than perfection.

Here both P1 and P2 can be considered to be representative of a typical brain's pyramid of patterns; however when the conclusion is fed back to the pyramid of patterns it is found that it does not fit in in any sensible way.

This is because the syllogism operation in an abstract logical processor acts upon the labels rather than the patterns. And so it is seen that if a mapping between two different systems is not identified as such but is instead treated as a direct connection then false conclusions can be reached.

Yet all too often other philosophical paradigms conflate the two and consider that abstract logical systems can act on words as though they are both simple strings of symbols and complex patterns. They do not acknowledge that a mapping is required in order to combine the two distinct systems of abstract logic and pattern identification logic. Since this mapping is not based on any rigorous logical system but rather upon a more haphazard system of trial and error, the deductive certainty of a theorem generated from an abstract system is lost once the theorem has been mapped onto the words that act as labels in the pyramid of patterns.

What this means is that logical arguments in the form of propositions and conclusions are not really logical as the patterns which are specified by the words of the propositions and conclusions cannot be treated as simple logical units and cannot be processed by a simple abstract logical processor.

So when a logical argument is presented as a series of logical steps and syllogisms it needs to be translated into a form that can be processed by a pattern identification system. This means that a domain for the data needs to be identified and perhaps also a range to the data and perhaps also a suggestion for a particular pattern or template that could be the best pattern to fit that data. (This book is representative of an argument of that form.)

Abstract systems are just that, abstract. They are entirely logical and being abstract they require no justification. They are defined by their axioms, symbols and rules of inference. There is no particular limitation on what the axioms, symbols or rules need to be. The only requirement is that they are sufficiently compatible with each other that a machine, albeit perhaps only a virtual machine, can be constructed that represents all the axioms and rules and which is capable of generating theorems. If an abstract system is incapable of generating theorems which are consistent with all its axioms and rules then it can be described as a null system.

Then the merits or otherwise of a non-null abstract system can be evaluated by how 'interesting' the theorems that it generates are. 'Interesting' in this instance means how useful the theorems are in being mapped onto patterns in the pyramid of patterns to create other insightful patterns.

It should also be noted that a simple logical processor, such as a computer, which is capable of generating theorems for a specific abstract system can only follow the rules which are specified by that particular abstract system.

So if for example a simple logical processor uses the string of symbols 'tree' as a trigger to perform a particular set of instructions, it does so only on that string of symbols, it has no connection to the pattern for which 'tree' is a label.

In contrast, should the logical processor which is used be a highly complex one, such as a brain, which does store the patterns to which various strings in the abstract system correspond, then that processor may well bring up the pattern associated with the string 'tree' when it encounters it while generating theorems for the abstract system. But in so doing it is leaving the deductive rigours of the abstract system and conflating the string of symbols with the pattern. And for people and brains, this may well be a natural (normative) thing to do.

Any philosophical argument that claims to be deductive in nature that leaves the domain of an abstract system of symbols and makes reference to the patterns (or concepts) associated with a string of symbols immediately loses the validity of their claim of deductive certainty.

For there are just two processes of drawing inferences (as indicated by Hume several centuries ago). 1. There are pattern identification processes whereby various templates are tested against the data and the best pattern selected. 2. Logical deductions from the axioms and rules of an abstract system. And that is all. Other pseudo-logical processes which conflate labels of patterns with symbolic strings of an abstract system can only lead to possibilities and not certainties, and those possibilities must be tested against the relevant data using a pattern identification system before a conclusion can be drawn. However typically the logical processor of both these systems is the brain and so conflation between the two can easily occur.

Communication from one person, even assuming it is an honest communication can only be an indication of the specific patterns that have been created in their brain. As such, for the receiver of the communication, it can only be considered to be a possibility for a pattern that might fit the data or the patterns in their own brain. In other words the communication can be labelled as an 'opinion' of the communicator.

This presents something of a dilemma for the receiver; however they have several possible choices:

1. Reject the opinion as being too incompatible with their own beliefs.
2. Accept the opinion as being a best pattern and adopt it as one of their own beliefs.
3. Expend considerable time and effort in gathering data relevant to the opinion and then using that opinion as a template in their own pattern-identifying system.

Which of these three is selected will depend upon the pragmatic situation that the person finds themselves in and the importance to which they attach the opinion.

Many people's lives are far too busy for option three and they will most likely accept options one or two. However for the philosopher option three is the only choice available to them. Possible beliefs have to be checked and tested against the available evidence before they can be accepted or rejected; it is not acceptable to be swayed by public opinion or peer pressure.

The reward is a well-tempered pyramid of patterns; one where all the patterns are arranged in an orderly way without schisms. Such a pyramid of patterns is far more harmonious and easy to use than one that is in disorder and with schisms. The cost is, of course, considerable expenditure of time and effort which may be hard to justify, remembering that the ultimate goal is happiness and good decision-making is essential for achieving happiness.

There may also be a middle ground between options two and three. If the original communication consists of not just one opinion but a whole raft of opinions and it appears that many other people also share these opinions (as one might experience in a learning situation at school or university) then there may be a compromise between options two and three. One can choose to test just one or two of the opinions (as for example in a school laboratory) and if it is found that they constitute the best patterns to fit to the relevant data, then one might feel confident in accepting all the opinions that have been presented.

So for example in the domain of astronomy and the solar system one's acceptance of all the opinions and theories might be bolstered by the identification of Saturn and its rings through a small telescope and then following its nightly progression across the sky relative to the stars. Hence one might feel confident in accepting the general canon of astronomical science.

Or alternatively there may be a compromise between options one and three. One could choose to test just one or two of the opinions and if it is found that they do not constitute the best patterns to fit the relevant data, then one might feel resigned to reject all of the associated opinions presented.

So for example in the domain of philosophy one might consider Kant's social imperative, Plato's promotion of slavery and Wittgenstein's "The world is made of facts not things" and decide that none of them are theorems of an abstract system nor the best fit patterns to the appropriate data and so one might feel motivated to reject the general canon of philosophical writings.

One of the tests for a good pattern, as opposed to a merely adequate one, is that it not only fits the data but is also capable of recreating the data. For overall it is a model of the world that is what one is trying to achieve, one that one can use to interact with the world effectively.

There is no requirement for a pattern to be 'falsifiable' or 'complete' or even 'true'; all that is required is that it is capable of reproducing the relevant data to a reasonable degree of accuracy.

Some possible patterns are considerably better at reproducing the relevant data than others. Take for example theories regarding the cause of lightning. The ancient theory that lightning occurs when the god of thunder casts one of his bolts has little to say about the physical constitution of lightning nor why lightning occurs most commonly during a storm. The more modern scientific theory is that lightning is a flow of electrons caused by a high voltage difference between the clouds and the ground which in turn is caused by the jostling of molecules in the clouds which causes some electrons to break free from their associated atoms and be carried to the ground with the rain which makes the ground negatively charged and leaves the clouds positively charged. The lightning strike then returns the electrons to the clouds with a great release of energy in the form of both light and sound.

While the former theory of lightning is simple the latter is considerably more complex, however the former is not capable of reproducing the relevant data at all whereas the latter has considerable power in reproducing the relevant data.

Sometimes two patterns are similar in their efficacy at fitting the data and the particular one selected will depend upon the particular environment over which it is to be applied.

Take for example the patterns or theories of Newtonian gravity as compared to Einstein's curved space. Both refer to the domain of masses interacting through space and time. Both provide useful models of the physical world. Einstein's curved space is more accurate for it can explain the precession of the orbit of mercury around the Sun to a greater degree of accuracy than can Newton's theory of gravity, but it is much more complex and harder to use requiring knowledge of tensor calculus. Newton's theory of gravity is not so accurate but is much easier to use and is sufficiently accurate for everyday use, such as calculating the height above the Earth of a geostationary satellite, that is a satellite that circles the Earth at the same rate as the Earth rotates.

So which of the two one selects will depend upon the use to which one wants to put it. Most people find the accuracy of Newtonian gravity to be adequate and the complexity of Einstein's curved space to be too great and so will select Newtonian gravity. However in the domain of cosmology and the physics of black holes the more accurate theory of curved space of Einstein will most likely be selected.

Once a pattern has been selected, it can then be added to one's pyramid of patterns.

At that stage it then constitutes 'belief' and can be used to interact with the world and to maximise one's happiness.

Once the pattern has been selected it is unlikely that its domain will be revisited with the intention of reprocessing the data to find the best pattern. It would take some significant data point for which the pattern did not fit for it to be deemed worthy that the pattern identification process for that domain to be revisited. For otherwise it would not be worthwhile to expend the time and effort of the brain's pattern identification processes.

And if one was communicating information about the domain of a particular pattern one might well claim that the pattern one believed was 'true' in order to distinguish it from other possible patterns that might be labelled as 'not true'. In other words it may well be useful for the purpose of communication to label a pattern that one believes as 'true'. So that is what truth is, at least according to the TTP paradigm. It is a label that is used to indicate that a particular pattern is believed and that that particular pattern is considered to be the best one for that particular domain.

This is in contrast to some other philosophical paradigms where it is considered that 'truth is a property of statements'. [Ref OCP] However according to TTP this definition is very weak as it makes no indication of how truth can be a property of a statement or of a communication, nor does it indicate how it can be determined whether any particular statement has this property or not. For without a definitive process for determining whether a statement has the property of truth or not, the claim that truth is the property of a statement is extremely weak. In practice of course the truth of a statement is adjudicated as an opinion of the communicator. And what is an opinion if it is not a communication of what one considers to be the best pattern to fit the relevant data? In which case the other philosophical paradigms might just as well hold that truth is an indication of the best pattern that fits the data which is of course essentially the definition of truth in the TTP paradigm.

Before leaving this chapter on logical systems I want to discuss a little further the quintessential abstract system of mathematics. I have asserted that mathematics can be modelled by an abstract logical system, yet the question can be asked: can all of mathematics be modelled in this way?

Certainly in its general use and also its use by scientists, it can be modelled in that way. This is amply demonstrated by the use of computers, which are essentially logical processors, which have proven themselves capable of manipulating the symbols of mathematics in an entirely logical and abstract way to generate theorems for that system. These theorems can subsequently be applied to the 'real' world of patterns in many different domains such as the counting of sheep or predicting the motion of a rocket in a gravitational field.

But there are other aspects of mathematics such as the question of whether there is an integer solution to the equation $a^3+b^3=c^3$? This question is equivalent to asking whether a string of symbols of this form, with 'whole numbers' in place of the letters, will ever be generated as a theorem of the system. Questions such as this are typically generated by taking the theorems that are generated by a mathematical processor and then feeding them back into a pattern identifying processor and looking for patterns among those theorems. Following this process, certain patterns can be identified and questions relating to those patterns, such as whether they can be extrapolated into new domains, can also be identified.

So for example, many integer solutions have been found to the equation $a^2+b^2=c^2$, and the question can then be asked can this also be extrapolated to the equation: $a^3+b^3=c^3$?

This is not so much a problem of mathematics as it is an exploration of the system of mathematics and as such it cannot be answered within the system of mathematics but instead requires a meta-system of mathematics, a system that is one step removed or above the system of mathematics and uses the theorems of mathematics among its many axioms.

Some mathematicians, such as Hilbert, have suggested that mathematics should be 1. Finitely definable. 2. Consistent. 3. Complete. However Gödel showed with the proof of his incompleteness theorem that such a description of mathematics was not logically possible.

However the abstract system model of mathematics as described in this book does not incorporate such illogicalities, in other words, it is consistent with Gödel's incompleteness theorem. For while the abstract system does require the system to be finitely describable (the logical processor which generates the theorems must necessarily incorporate all the axioms and rules of the system and so necessarily they must be finite), it does not require the system to be either consistent or complete.

Indeed for the abstract system model the only requirement for consistency is that the axioms and rules are not so contradictory that the logical processor cannot be constructed which incorporates all the axioms and rules. Further, the theorems of an abstract system are, on their own, meaningless strings of symbols, there is no process which could identify a contradiction among the theorems.

For example, there could be an abstract system, not mathematics, which generates as a theorem the string '2+2=4' and the string '2+2=5'. There would not be any indication of inconsistency as they are just strings of symbols. The only possible identification of a contradiction would be encountered by an interpretation of such a string following a mapping process onto the pyramid of patterns. Clearly there would be a problem in mapping the symbols of '2+2=5' onto the patterns for 'two', 'plus', 'equals' and 'five'. But this would be a problem with the mapping and not with the abstract logical system itself. And since the string '2+2=5' is not considered to be a theorem of a mathematical abstract system, there is no problem with the system of mathematics.

And as for completeness, there is no requirement for an abstract system to be 'complete'. The system is defined by its axioms and its rules and the number of theorems that can be generated may be finite or they may be infinite, but either way there is no requirement of completeness.

Another aspect of mathematics as an abstract system is a question of how the axioms and rules are created in the first place. Clearly they cannot be generated by the system itself nor by any strictly logical process. The only way that they can be created is by a brain with a pyramid of patterns playing around with various logical possibilities on a trial and error basis and looking for a system that generates interesting theorems. And that is what some mathematicians do; they search for symbols and rules that are compatible with the pre-existing symbols and rules and which can expand the domain of mathematics. The invention of calculus by Newton and Leibniz is one example of how the domain of mathematics was expanded.

The invention of 'i' as a solution to the equation '$z=\sqrt{-1}$' is another example. Both inventions and other additions to the system of mathematics have generated many interesting theorems. Part of the cover of this book is an example of this, it being a part of a variation of a Mandelbrot set which incorporates the use of $\sqrt{-1}$. The two-dimensional assortment of coloured pixels can be considered to be a theorem of the system.

CHAPTER 9

Pyramid of Patterns

One of the purposes of studying and exploring philosophy is to develop a well-tempered pyramid of patterns, one where one's model of the world is not only accurate and useful but also one where there are no gaping holes or non sequiturs; one where a conglomeration of disparate facts are smoothed out by simple patterns into a well-structured and unified whole.

When a mind encounters new data it may choose to ignore it as being uninteresting (for example perhaps the sight of a snail on a leaf) or alternatively it may be acted upon if it is considered that the new data is sufficiently interesting and significant (for example perhaps the sight of a tiger behind a tree). Also the new data can be evaluated to see whether it fits in well with one's pyramid of patterns (for example perhaps the sight of a dog behind a gate) or whether it does not (for example perhaps the sight of a dolphin driving a car).

If the new data does not fit well with one's pyramid of patterns then it must be decided whether the data is sufficiently significant to warrant a re-evaluation of the relevant pattern (for example one might hold the theory that dolphins are usually found swimming in the ocean). The first thing to do is to evaluate the accuracy of the new data (was it actually a dolphin in the car or merely a person dressed in a dolphin costume?). If the data is evaluated as being accurate then it is necessary to assess whether it is important enough or significant enough to warrant a reassessment of the relevant pattern. The data could merely be ignored as being a weird anomaly that does not impact on the general validity of the already held relevant pattern (for example it may be held that dolphins do usually live in the sea and perhaps there is just one uniquely trained dolphin that can drive a car). Or alternatively it could be decided that, pending additional information, a re-evaluation of the relevant pattern is required (which could lead to the general idea that dolphins live in the sea and can drive cars). The decisions made during this process will depend somewhat upon the particular circumstances and character of the person making them.

Also it is quite possible that the person might seek advice from others as to whether they have experienced similar data in order to determine whether a re-examination of the relevant pattern would be merited.

And if the original pattern was not so much derived from underlying data from personal experience as it was adopted from a communication and associated peer pressure, then it would not be possible to simply re-evaluate all the data for oneself. In such a case either the new data point will simply be rejected as unreliable and the original pattern maintained, or there will need to be a complete rejection of the adopted pattern and new relevant data collected and subsequently a new pattern identification process would be implemented.

Which option is chosen will depend on the character and situation of the person involved.

For example, if the original pattern was: 'The emperor is but one step removed from a god' and the new data was the sight of the emperor torturing a kitten, then if the person benefitted greatly from the imperial system they might be inclined to overlook the non-fitting data, whereas if they felt oppressed by the imperial system they might look to undertake a complete revision of the deific qualities of the emperor.

These examples may demonstrate something removed from the

original contention that the brain seeks to find the best pattern to fit the data, and certainly this is the case for patterns at the lower and more fundamental part of the pyramid of patterns, at the higher end the brain's prime motivation of seeking happiness comes into play and possibly in an opposing way.

So if the person considers that they would be happiest by maintaining the status quo and supporting the emperor then they will seek to find justification for dismissing the new data as perhaps nothing more than an anomaly. And by doing so they will maintain their belief in the deific qualities of the emperor.

On the other hand if the person is not so enamoured of the emperor but nevertheless had hitherto believed in the deific qualities of the emperor, they may consider that it would maximise their happiness to undertake a complete revision of the relevant data and associated patterns.

This could result in what could be called a minor paradigm shift. For if the result of the investigation is that the emperor has no deific qualities and most likely never did, then all the patterns in the person's pyramid of patterns that lie above the pattern ' the emperor has deific qualities' and use that particular pattern as input data to create high-level patterns such as : 'I am entirely safe and blessed to be living in this imperial state' will need to be re-evaluated as well.

This paradigm shift process can cause a certain degree of 'stress' to the person as the beliefs that once were held as true change and mutate. For this reason it is not a process to be undertaken lightly. The rewards for the improved and well-tempered pyramid of patterns need to be offset against the time and effort required to achieve this.

Paradigm shifts can also occur in academic subjects such as physics and philosophy. A well-known one being the Copernican shifts from a geocentric (Earth-centred) solar system to a heliocentric one (Sun centred). There was considerable opposition to that shift, particularly from the Catholic Church, which considered that such a shift was a threat to their own God-based cosmology.

While one would expect in academic circles that it is always the best pattern that will be searched for and adopted without reference to any factors of happiness nor time constraints, it must be remembered that academics are people too and are subject to constraints of happiness and time. They are also subject to peer pressure and so it may be that the best pattern is not necessarily the one that is adopted, even in the best academic circles.

What this all means is that it can occur that people will choose not to adopt the best pattern but prefer to adopt some other pattern which they believe will bring them the most happiness in their lifetime. Once the decision is made to remain with the status quo the reasons behind this decision will be lost and forgotten. What remains is their own self belief in the integrity of their pyramid of patterns.

In science there is a high degree of objectivity. For given all the writings and theories of science it should be possible for a student to recreate the experiments or observations and follow the processes of inference from the basic data to the most advanced theories and test the accuracy and validity of the established theories along the way. And at each step the input data can be tested to see whether the established theory is indeed the best pattern that fits the data.

In philosophy however the degree of objectivity is much less than in science. For there is no direct and simple linkage between the theories of philosophy and the raw data from the senses in the way that there is for the theories of science. Even for the philosopher who arrives at a particular theory, the actual processes by which they arrived at their conclusion may not be entirely clear to them. Typically, their theories can be labelled as 'opinions'. A personal opinion is created from the sum total of their experiences and the patterns that they have created along the way and is based upon the data they have encountered. This will undoubtedly include some less-than-ideal patterns and some higher level patterns based upon those less-than-ideal patterns but which nevertheless they have labelled as 'true'. Some of these will undoubtedly have come from things they have learned from other people and from the culture they were raised in.

The expression of an opinion is also a consequence of the action of speaking or writing and so will be subject to what they think will bring their happiness; perhaps by impressing other people or influencing their thinking and as well as what they consider to be an honest contribution to the general canon of philosophy.

So the line of logic and reasoning that leads from basic sense data to an opinion is complex and one that may be hard, if not impossible, to reveal. For of course any theory of philosophy that relates to the real world or more accurately to one's model of the real world must ultimately be based upon patterns created from sense data; if a philosophical theory is not so based then it is indistinguishable from fantasy and fiction. And of course if a philosophical theory does not relate to a model of the real world such as in mathematics or logical processes then there is no requirement for it to be linked to patterns based on sense data.

Philosophy is full of opinions. Yet in order to make sense of philosophy and by implication to make sense of the world, it is necessary to focus on the facts and not on the opinions.

A good and lasting philosophy must be based on indubitable facts and proceed using lines of reasoning that can be followed by other people. When assumptions are made, which they clearly must be, these should be clearly identified so that they can be revisited and revised should this seem beneficial. It is important that the assumptions should be identified as each new assumption leads to a new and different paradigm.

Philosophical theories should be treated as opinions or merely possibilities for efficient theories until such time as they have checked out and a logical sequence identified between raw sense data and the theory.

Other philosophical paradigms tout 'logic' as being at the heart of their philosophy, yet it is rarely actually used as a process of inference to arrive at an interesting conclusion. This is because, as has been previously pointed out, there is a logical gap between an abstract system such as pure logic and the system of words and patterns. This gap can only be bridged through a mapping process, one that is based upon trial and error rather than logic.

This pseudo-logical system typically incorporates syllogisms, propositions and conclusions. It portends to act upon words and moreover the patterns or concepts behind those words.

The presumption that patterns can be treated as simple units that can be manipulated by a logical process to generate a valid conclusion (one that necessarily follows from the propositions) is flawed. It is flawed because patterns can only be treated logically in a system that contains information on the nature of the patterns and the only known system that contains this information is a brain. So it is only a brain that is capable of manipulating patterns in any sort of logical form. And this is quite different from a logical process of say a syllogism.

At best, the simple manipulation of words in syllogisms can lead to possibilities. But then these possibilities can only be believed if they constitute the best pattern to fit the relevant data.

Sometimes of course logical inferences can have the appearance of being valid. Consider for example the often quoted classical case:

P1   Socrates is a man.
P2   All men are mortal.
C   Socrates is mortal.

To evaluate those using the process of the best fit to the data one could infer that:

P1   Fits the relevant data.
P2   Fits the relevant data.
C   Fits the relevant data.

None of these statements seem to be particularly interesting in that one could just as easily have started with the conclusion as with the propositions.

Consider another example with a similar grammatical structure:

P1   A peanut butter sandwich is better than nothing.
P2   Nothing is better than perfection.
C   A peanut butter sandwich is better than perfection.

Evaluating these statements with a view to finding the best fit to the data:

P1   Fits the relevant data.
P2   Fits the relevant data.
C   Doesn't fit the relevant data at all.

The grammatical construction of this syllogism is very similar to the previous one and if one was treating the words as simple strings of symbols which follow the syntax of the English language the conclusions would be equally valid. Yet they are not. This is because the words (or labels for patterns) are treated as simple logical units rather than the complex patterns which they denote.

Some philosophers might try to explain that the inferences which are not valid are anomalies and are a consequence of the unusual use of particular words. But without a specific logical system of determining what constitutes an anomaly and what does not, the whole logical integrity of the system fails. And since the system fails it cannot be used as a process of inference within a rigorous and well-defined philosophy. And this is why this form of logical inference has no part in the Pattern Paradigm philosophy.

It is the ideas of philosophy that are important; the ideas that constitute the best patterns to fit the data. And within the domain of philosophy one is looking for the deepest patterns that fit the most general of data. Words and language are merely the medium by which such ideas can be communicated.

It may be useful to identify the different levels of interpretation of a word in the English language. Take the word 'table' for example.

Level 1: At the most fundamental level this word constitutes a string of symbols: 't', 'a', 'b', 'l' and 'e'. At this level the words or symbols can be processed in a simple logic machine. For example the Enigma coding machine could translate the string into another string such as 'a', 'z', 'z', 'f' and 'g'.

Level 2. At this level the individual symbols are combined together to create a word. This word can be compared with other words in the dictionary. If the information in the dictionary is included in the axioms of an abstract system then theorems can be produced of the form: 'a table is an item of furniture'. At this level the word has no meaning neither do the theorems generated from the abstract system.

Level 3: at this level the word is treated as a label for a pattern, this requires a logical processor with a pyramid of patterns, most likely a brain. In this way the word can be linked to a pattern. Thus the word 'table' can be associated with all patterns which have been labelled with 'table', such as 'the table in front of me', 'logarithm tables', 'the Round Table', 'glass tables' and so on.

Level 3b: this level is just a variation of level 3. It is one where the person either has no knowledge of the pattern identification process or perhaps no knowledge of the distinction between phenomenon and noumenon or perhaps just chooses to ignore it. In this case instead of the word acting as a label for a pattern, the whole pattern identification process is ignored and the word instead is used as a label for a 'real world' object. Or it could be seen as a convenient abbreviation. So instead of saying something like "the pattern that I have created from sense data which I've labelled a 'table' and which I have no doubt that you have also identified as a 'table'", one could abbreviate this simply to 'table'.

All that a person knows about the world comes from pattern identification of data from the senses. Even the hypothesis that there is a real world out there is no more than the best fit pattern to the relevant data.

So where a person might say "there is a table in front of me", a philosopher might translate this into something like "according to my model of the world and the pattern which best fits the sense data I am receiving I can conclude that there is a table in front of me."

And so the question "what is?" can be translated into the form: "What is the best pattern that fits the relevant data?" So for example instead of asking "Does A cause B?" The question can be translated into: "Does the pattern of A causing B constitute the best pattern to fit the relevant data?"

Some of the conundrums of philosophy such as 'The Ship of Theseus', and 'Does a tree falling in a forest make a sound if there is no one there to hear it" can be unravelled using this approach. So instead of asking 'What is...?' One can ask 'What is perceived...' and then searching for the best pattern to fit the relevant data.

It may be useful to look at how a pattern is stored in a logical device such as a brain. It would seem likely that it is stored as a code that connects sensory input to motor output. It is stored as a code in the same way that data in a computer is stored in a coded form. For example if one were to read the data directly off a music compact disc (cd), it would appear as nothing more than a long string of 0's and 1's. But the CD player that plays the CD has information on the form of the code and can reconstruct long strings of 0's and 1's into motor output instructions which can then be sent to the loudspeakers which then enables people to hear the variations in air pressure which they generate as music.

It is only brains that have the ability to store information on patterns (that have been identified from sense data), so it is only brains that can process information obtained from sense data. Thought and logical inferences need a logical process for those inferences, for without a logical process the conclusions would be indistinguishable from entirely random ones.

Every logical process requires a logical processor. For simple symbolic manipulation a pocket calculator or computer is sufficient. For the processing of patterns a logical processor that has access to a pyramid of patterns is required and so far it is only a brain that has that.

It may also be useful to reiterate that all one knows of the world is a model created from patterns. Words are labels for those patterns and truth is a label used to indicate a degree of confidence in the efficacy and accuracy of a particular pattern.

There may be a temptation to consider that words correlate directly with a noumenal type real world in what is generally termed 'naïve reality'. Certainly this idea could be considered to be a pattern that fits the data and one that is useful for everyday life. However it is something of an oversimplification when it comes to philosophy and the understanding of the relationship between words and a hypothesised noumenal reality. It is similar to the relationship between Newton's theory of gravity and Einstein's theories of curved space-time. While Newton's theories are fine for everyday use and are certainly simple to use, one needs to turn to Einstein's theories to obtain a better understanding of forces and space and time.

A more accurate path for the relationship between words and the noumenal world is that words are labels the patterns and patterns are identified from sense data and that they sense data originates from some hypothesised noumenal world.

There is also the temptation to claim that the pattern is one has subjectively labelled as 'true', because they seem to be as good can reasonably be expected, are also 'true' for all other people. However this temptation can be moderated by recalling that all one knows of the world is a model and everyone creates their own model. It may be that the pattern fits the data so well that it is considered to be indubitable, but this does not preclude the possibility that a better pattern might one day be identified. Perhaps the most indubitable patterns that can be made is the one that corresponds to 'I am', the one that identifies oneself as the perceiver of the world and the creator of patterns.

The method of communicating a pattern from one person to another is to communicate a number of relevant data points or a domain for a set of data points (if it is considered that the receiver of the communication already has most of the data points stored in their pyramid of patterns) and then to specify the pattern or patterns that fit those data. It is then up to the receiver of the communication to determine whether the suggested pattern does indeed fit the data or not and whether to adopt that pattern as one of their own beliefs or not.

For example, in a court of law a prosecutor might start with the pattern 'the defendant is guilty' and then proceed to identify various relevant data points which fit in with this pattern. The defending attorney is likely to start with the proposed 'My client is innocent' and then proceed to cast doubt on the data points of the prosecuting attorney and to introduce other data points which fit in with his pattern of innocence. Subsequently the jury can decide for themselves what they consider to be the best pattern to fit the relevant data.

A similar process occurs in philosophy. People interested in philosophy propose patterns such as 'truth is a property of statements' or 'it is possible to have secure knowledge of what is right and wrong' and then identify data points which they consider support their pattern or theory. Other people may well propose different patterns and cast doubt on other people's data points and introduce different data points of their own.

This book is of that form. I have identified a number of patterns, some of which I refer to as anchor points, and have endeavoured to show how they can fit many of the normative data points that most people will encounter in the world and thus to gain a simple and comprehensive understanding of the world.

The process for communicating how a particular theorem of an abstract system can be generated is quite different. Abstract systems have no connection to sense data and their axioms and rules of inference are overt and known. So all that is required to communicate how a particular theorem may be generated is to specify an algorithm that sets out how the particular rules of the system are to be applied to a particular axiomatic starting point. This is commonly referred to as a proof.

For example if one were wishing to communicate how a particular configuration of Conway's game of life can be arrived at, all that would be needed to be communicated would be for the initial configuration and the number of cycles of the system that must be completed in order to arrive at that particular configuration.

Chapter 10

Decision Making

Broadly speaking brain processes can be divided into two main parts. The first is the construction of a model of the world by creating patterns from sense data. These patterns can then be stored as memories and knowledge. The second part is the use of these patterns to make decisions and the sending of instructions to motor outputs in order to interact effectively with the world and gain happiness.

In this chapter I want to focus on the second part and in particular the interactions with other people.

Humans are a social species and interactions with others are a major part of brain activity and decisions. The world is a complicated place and learning from others and teaching others constitute an essential part of the human experience.

Everyone makes decisions that they believe will maximise their happiness. So if one wanted to assist other people in making decisions one would explore with them the possible choices they might have, together with their short and long-term consequences and how those consequences might impact on other people's happiness. And then allow the other person to make their choice.

However it must be remembered that the person giving the advice has the motivation of improving their own happiness as their prime motive for giving the advice. So what can happen is that instead of exploring possible choices, consequences and resultant happiness of the other person, the one giving the advice may try to push the other person into making a particular choice which is more to the benefit of the one giving the advice than the one receiving it.

So for example they might say 'you should do A', 'A is the right thing to do', 'only bad people do B', 'It is wrong to do B'. Clearly they are trying to push the other person in the direction of doing A and not doing B. And certainly if the one with the decision to make is unsure of the possible consequences of A or B, then this could be a strong influence on their decision making in favour of option A and opposed to option B.

Perhaps the best way of dealing with such advice and to put it in perspective is for the decision maker to translate the statements of the advice giver into a more accurate and meaningful form. Thus if the advice giver says 'you should do A', the decision maker can translate it into something like: 'the advice giver would like me to do A as the advice giver considers that this will maximise the advice giver's happiness'. And then the decision maker can take what the advice giver has said into account when making their decision.

This brings us to the domain of morality and the theory of morality, otherwise known as ethics.

Morals and moral codes of behaviour stem from pragmatic concerns of living within a community and are typically determined by a general consensus of opinion and putting into practice the best codes of behaviour that are considered to benefit the community. Thus for example if a community considers that spitting in the street is inappropriate behaviour and the people of that community generally refrain from doing so and have disdain for anyone who does so, any visitor to the community can observe this behaviour and infer that spitting in the street is opposed to the moral code of that community.

A country's legal system is distinct from its moral codes though undoubtedly they are linked. A country's legal system is set out by the government and includes both justice and penal systems.

The legal and penal code of a country can have a major influence on the consequences of a decision, particularly if the action that follows from the decision is considered to be in violation of the legal code. Factors that might be considered include the likelihood of being identified as a violator of the code, the likelihood of being apprehended and convicted and the likely consequences of conviction, and especially the unhappiness that this would be likely to bring. This unhappiness would have to be compared with the expected happiness that doing the original action would bring

Both moral codes and legal codes may need to be taken into account when considering the possible consequences of a particular action.

When wishing to influence other people's behaviour, moral assertions can be considered to be a simplification of an analysis of the consequences of a particular course of action and in particular its long-term consequences. Then the claim that another person's actions or proposed actions are 'unethical', ' immoral' or 'wrong' can be taken as a warning of potential dire long-term consequences. But this will not necessarily influence the other person's behaviour if they don't regard the other person's opinions very highly or consider that the other person's analysis of the situation is inaccurate.

It would likely be more effective if the potential consequences, particularly the long-term ones, were stated explicitly including how those consequences are likely to impact on the happiness of the person whose behaviour is under consideration. For if it can be shown that the person's proposed actions will most likely lead to unhappiness for that person and if that person can be persuaded of this and that there is another course of action that will bring them greater happiness then naturally they will avoid the behaviour that would likely bring them unhappiness.

However it must be emphasised that this can only be effected if the decision maker believes this and it makes sense for them, given their individual character and circumstances. It is not sufficient for the person giving advice to convince themselves that if they were in that situation that they would make a particular action as they consider it would bring them the most happiness. It has to be the assessment of the person making the decision.

The idea, as is put forward by other philosophical paradigms, that there is one universal ethical standard that applies to all people is perhaps best described as naive as it does not take into account the particular circumstances of people nor their particular characteristics nor the logic of the decision making processes. These attempts by those other paradigms to use ethics as a means of influencing other people's behaviour can best be described as 'propaganda'. It can be described as propaganda because it is intended to influence the behaviour of the people without due consideration of what is best for them. A more effective ethical strategy would be one which teaches people how to evaluate the long-term consequences of their actions and how those consequences would impact on their personal happiness.

Certainly this strategy of propaganda can be used for influencing the behaviour of others but it is certainly not the only strategy and probably not the best either.

A popular strategy is to complain to other people about their behaviour or attitudes and claim that they are 'wrong' or 'immoral', but this may not necessarily be particularly effective. This is because other people will only change their behaviour if they perceive that it is in their best interests to do so; in other words they will only change if they believe that it will increase their happiness.

People are influenced by social pressure and peer pressure. It is natural for people to want a feeling of security for being part of a group. This can lead to what could appear to be altruistic behaviour, where an individual might risk personal loss in order to achieve the approval of the community.

People make their decisions based upon what they consider will make them happy and this is quite distinct from what might actually make them happy and from what other people think might make them happy. Personal decisions are based upon the person's particular characteristics, situation and perception of the world.

However the strategy of propaganda will only work upon people who feel a strong tie to the community and who feel that doing what others consider to be right is important to them. For other people, propaganda may have little or no influence on their decisions. For these people a strategy of exploring the long-term consequences of their decisions and actions and showing how these might impact upon other people and how those impacts might return to influence their own happiness might well be a more effective strategy for influencing their behaviour.

People like to cooperate with others as this brings many benefits to the community and hence to themselves. It also provides security and a sense of well-being. In other words people cooperate because they can perceive the benefits and believe that it will bring them happiness.

Chapter 11

Making better sense of the world

The Pattern Paradigm described in this book constitutes a model for making sense of the world. Undoubtedly there are gaps in the description of the model and the model itself is nothing more than a model, but this just means that there is some filling in of the details that is required rather than any major or fundamental restructuring of the model.

I have kept the description fairly lean as I only want to describe the innermost skeleton of the model. I don't want to compromise its integrity by making assumptions that only apply to some people but not to all. I have not included any complex or convoluted arguments as this would only reduce the efficacy of the description of the model. Instead I have merely pointed out the links between the different facets of the model and I leave it up to the reader to flesh out the skeleton from their own particular environment, knowledge and situation.

I have shown using just a few reasonable and justified anchor points how a framework can be constructed which can then be used for generating an efficient, accurate and self-consistent model of the world.

Philosophy is all about making sense of the world; creating a harmonious pyramid of patterns that is sufficiently comprehensive to be able to relate in some broad way to the full range of known facts. It is about creating a comprehensive model of the world that can be used to facilitate interactions with the world. There are benefits to having a better understanding of the world, not only for having a more harmonious pyramid of patterns but also as a means for a person to make better decisions and hence have a better chance of realising their ambitions and achieving happiness.

The Pattern Paradigm is only a model, for there is an impenetrable barrier between the data that we receive through our senses and the actual source of that data, hence the best that can be achieved with regard to knowledge of the world is knowledge of a model of the world. The model of the world that we create is all that we know of the world and so it is natural to believe that it is real. For most purposes the belief that it is real is sufficient for interacting effectively with the world; however for philosophers the identification of this distinction is essential for the formation of an effective and efficient philosophy. Even so, many other philosophical paradigms fail to make this distinction and consider that their paradigm is the truth and that it relates directly to an objective picture of the world rather than relating to a subjective model of the world.

So what one arrives at is a philosophical paradigm, it is not the final word on philosophy nor on making sense of the world, but it is one that enables one to interact effectively with the world and to make good decisions.

The philosophical paradigm described in this book which I have referred to as 'The Pattern Paradigm' is one that maximises the requirements of accuracy, simplicity and comprehensiveness.

It is accurate because the patterns herein described constitute a good fit to the data regarding the world. It is simple because it only uses simple logic in simple language and makes only a few assumptions all of which are explicitly identified as being assumptions (anchor points) and which can be considered to be the best patterns that fit the data. It is comprehensive because it covers most of the major areas of philosophy.

And while it makes the claim to be comprehensive it does not claim to be complete. There are many gaps and interesting areas that it does not cover. This is in part because it only describes a basic framework.

However, just as there are aspects of the pattern paradigm that do not fit comfortably with other philosophical paradigms, there are aspects of other philosophical paradigms that do not fit comfortably with the pattern paradigm; I have touched on some of these elsewhere in the book.

The Pattern Paradigm is not a description of the way the world should be or could be, instead it is a model of the way the world is. However, given its original approach both in method and content, when compared to other philosophical paradigms it may require a paradigm shift for other people to appreciate it.

It is important for anyone and everyone to have a good understanding of who they are; hence the importance of the pattern paradigm. As a relevant analogy, it would be entirely inappropriate for an elephant to consider that she was a giraffe and to exhibit all the behaviours of a typical giraffe. So too it would be inappropriate for a person to think that they are someone whom they are not.

The true nature of humans cannot be hidden, nor can it be altered.

Philosophy is a discipline of communication, for if there is no communication there can be no philosophy. But philosophy itself is not to be found in the writings of philosophers. The writings contain a description of the philosophical ideas, but it is only when those ideas are communicated and replicated in the mind of the receiver that the communication of philosophy is achieved.

Also in the domain of communicating ideas there are many ideas or theories that are not the 'best pattern to fit the data'. There are many ideas that are put about as entertainment or fiction - ideas that are imaginative and could possibly fit the data but which do not actually fit the known data, for example stories regarding Robin Hood. There are also ideas which are based upon wishful thinking but which again do not fit the data, for example theories about life after death.

One of the uses of TPP is that its methods can be used to evaluate the accuracy of such ideas for if they do not constitute the best pattern to fit the available data then they can be rejected and discarded or at best retained as possibilities pending further data acquisition.

This is especially useful in the domain of philosophy where many ideas have been put forward with determination and an air of certainty but which are not entirely justified. They are often presented without clear statements about what the assumptions are nor having a clear and logical process for inferring theorems from those assumptions. Instead they implicitly assume that normative beliefs are somehow true and make inferences from these beliefs using some vague process of inference. It is akin to assuming that the world is flat and then inferring that it must be supported on the back of a giant turtle. With few facts available, as was the case many thousands of years ago, such a conclusion would not have been entirely unreasonable; but with the facts available now such an assumption and conclusion are nothing short of ludicrous.

Yet when new facts come along many people are extremely reluctant to abandon their earlier beliefs. They have created patterns associated with those beliefs which makes sense to them and they see no reason to upgrade them.

It is the same with philosophy, politics and religion; people are reluctant to abandon their beliefs even if they are as far-fetched as believing in a flat Earth. Of course they have every right to do so, particularly if they do not consider that changing their beliefs will improve their happiness.

However for the purpose of achieving a comprehensive and harmonious pyramid of patterns, it is important to be prepared to let go past beliefs in favour of new and improved evaluations of new data.

The human brain is a complex organ which follows complex logic, but this does not mean that it is illogical. It just means that the logic is complex and cannot always be followed or understood by an outsider.

Philosophy would be nothing if it were not believed and yet many philosophers have ignored this obvious fact and have instead claimed that the statements of philosophy are objectively true and indubitable, irrespective of whether people believe them or not. Such claims can be considered to be communications and may need to be translated so that they make better sense. For example if person A claims that 'X is true', one might need to translate it to 'A considers that X is the best pattern that fits the relevant data' or 'A considers that X is a theorem of some, possibly ill-defined, abstract system'.

Claims of truth regarding the world inevitably rely on assumptions and often those assumptions are assumed and implicit rather than stated and explicit. Also it is also the case in philosophy that claims of truth are often based on a faulty logic that conflates pure logic with language.

The best way out of the conundrum of what it means to say that a statement is 'true' is to claim that statements are a representation of an idea or ideas and that if the person believes the idea then they are making a claim that such a statement is a representation of those ideas which are true for that person and hence they can label the statement as being 'true'.

It is the same for statements one makes oneself. While one may believe them to be true and have no doubt that they are 'true', one also needs to accept that they are based on assumptions and a pattern identifying logic. So, even as an honest representation of one's beliefs, it is best to be considered as a communication from one person to another.

When one delves into the logic of the brain one finds it is not simple and straightforward. At the deepest level it is simple, the logical processes of the brain seek to find the best patterns that fit the sense data and in so doing create a model or picture of the world and one which is in essence common to all people. However at the highest level, the fact that the brain seeks happiness as its primary purpose suggests it is possible that the simple process of searching for the best pattern can be skewed.

One example of this is related to a paradigm shift and a person's reluctance to adopt it. Even though a new paradigm may be an obvious better fit to the available and relevant data than the old one, a person will be reluctant to accept it unless they believe that it will bring them greater happiness. If they do not believe that it will bring them happiness they may well attempt to concoct arguments to justify their non-acceptance of it.

It could be argued that what makes a person happy is at the root of humanity. It has evolved over many hundreds of millions of years of genetic variation and reproduction. In this respect it was created as a consequence of a totally blind process; there was never any goal or intention. Different genetic variations generated different forms and intensity of happiness, some of those genetic variations survived to reproduce successfully while others did not.

It is this goal of happiness that motivates people to seek food and shelter and even to do philosophy. It would seem that a necessary requirement for achieving happiness is to have a good model of the world so that one can make good decisions and efficiently meet one's physical needs, and perhaps this is the case for the basic essentials of a model of the world. However at the highest levels of a person's pyramid of patterns this is not necessarily the case. For at the highest levels there is not a lot of relevant data that might pertain to a particular pattern. In other words, the selection of a particular pattern may be somewhat arbitrary as the data is scant and perhaps not even reliable. In this scenario it is possible that a pattern is selected with a bias towards the expected happiness such a pattern will bring rather than as a strict finding of the best pattern to fit the data. So, for example, a person might choose to believe that they are the best motor vehicle driver in their city despite the evidence for this being extremely scant. They do this because they consider that it will bring them self-esteem and a sense of well-being and happiness.

That which makes a person happy depends upon their situation and their character. Notably it is not a part of the pyramid of patterns. It is related in part to physical comfort and also to mental comfort and is most likely set by one's genes and hence is unalterable. Schopenhauer put it well when he said: "Man can do what he wills, but he cannot will what he wills." Physical comfort is fairly obviously warmth, freedom from physical pain, general health and so on. Mental comfort is more to do with feelings of security, achievement, freedom, friendship and a sense of community.

While no rational person would deny the evidence from their eyes should a tree fall in front of them, they might well deny that the Earth goes around the Sun, despite the considerable amount of evidence for it. This inclination to deny the obvious evidence is noticeably prevalent when people have held a particular belief for most of their lives and are then confronted with evidence to the contrary. They may simply refuse to accept the evidence or refuse to accept its significance; this is because they do not consider that their happiness will be enhanced by accepting the new knowledge or paradigm.

Some beliefs are so deeply ingrained that some people might choose death rather than change their beliefs. This may be in part because a person's beliefs define who they are and to change one's deepest beliefs can be very challenging and scary.

An example of this occurred when people survived a plane crash in a remote part of the Andes Mountains in 1972 only to face starvation when rescuers couldn't locate them. The clear solution to the problem was for the survivors to eat their dead compatriots. (The crash and was above the snow line so the bodies of those who perished in the crash itself were well preserved.) And this is what most of the survivors did. However some of the survivors would not do this as the taboo of eating human flesh was too powerful for them to overcome even though they knew the consequences would be their own death; some of them even going so far as telling the other survivors that they would be welcome to eat their body after they died. And so that is what happened. The story is recounted in the book 'Alive' by Piers Paul Read.

So too with religion, many people would risk death rather than change their religious beliefs. This may also be influenced by their belief in an afterlife and the fear that if they change their religion then they will face an afterlife of damnation in hell rather than in the bliss of heaven. Or to put it another way they fear that their future happiness will be greatly reduced if they change their beliefs. A notable example of this is Sir Thomas More who refused to accept King Henry VIII as the head of the English Church and chose death together with the expectation of a blissful afterlife rather than change his beliefs and live but risk eternal damnation. He was beheaded in 1535.

This could be considered to be surprising since there is no verifiable data which upon analysis leads to a best pattern for the existence of a God. Neither is there any data which upon analysis is the best pattern for the non-existence of a God. All the popular evidence for a God and an afterlife is based upon unverifiable stories and hearsay, which could be seen as strange considering the huge influence that religion plays in so many people's lives. However this does not mean that the practice of religion does not bring benefits to its practitioners, in fact the empirical evidence could suggest that it does.

People cannot live without a certain amount of tradition and culture. If people were brought up without culture they would not know how to behave within a harmonious society and would have to work it all out from first principles or more likely simply by trial and error. This would be an uncomfortable process and would be something akin to the anarchic situation described in William Golding's fictional account of youths stranded on an island without any adults and the chaos that ensued in his book 'Lord of the flies'.

It is not just humans that need culture. An article in New Scientist in 1996 describes how young bull elephants who had been orphaned from an early age after their parents were culled and were then released into a different wildlife reserve ran amok attacking tourists and killing rhinos. It seemed that a lack of discipline from older animals turned the young elephants into delinquents.

People need culture in order to have some grasp on how to live harmoniously with other people. The so-called Golden Rule dates back to antiquity and states something along the lines of 'Treat others as you would like to be treated yourself' and is a good place to start. It is often associated with the complementary Silver Rule which states something like 'Do not treat others in a way that you would not want to be treated yourself'. In effect what they are saying is 'Be nice', because then, hopefully, other people will be nice back to you and we can all live in harmony. While the 'cost' of being nice to other people may not always be directly repaid by the happiness gained from other people being nice back to you, it is repaid by living in a harmonious community with all the benefits that that brings. It is certainly a much better strategy than being unpleasant to other people as that unpleasantness will inevitably and eventually rebound. The pleasure one might gain from hitting someone that one does not like is inevitably more than offset by the pain one would experience when the person hits you in retaliation. The net pleasure less pain would be a lot of pain.

Despite many centuries of philosophical writing no one has managed to present any moral law that is more specific than the Golden Rule and to which there is general agreement. This is most likely because of the impossibility of making any specific universal law that does not restrict the freedom of human nature. Those that have tried and claim that people 'should do A' or 'should not do B' are merely expressing their own personal opinion of what works for them in their particular social and political environment.

The way we live now in the modern world is very different from that of our ancestors just 5000 years ago. It requires different skills and a different culture. It is not just a different technology and the necessity of adjusting to living in close proximity with strangers in urban centres that makes modern life so different as it is the invention of writing and its huge influence on civilisation.

No longer is communication between people limited to just one time and one locality; with writing we have the ability to communicate across time and space and to more people than just those within earshot. For example the writings of ancient Greek philosophers can be read by people thousands of years later and thousands of kilometres away. Also what is written down can be read over and over and can be analysed as to content and significance.

For some people what was written down took on greater significance than just the words of the communication. Promoters of religion considered their writings of their scriptures were somehow 'sacred' or 'holy'. Christianity has the Bible, Islam the Koran and Judaism has the Torah and the Talmud. One can hypothesize that this was done to promote the religion and to create the illusion that somehow writings are more than mere communications but are 'truths' and hence to deflect any questions regarding the validity and the importance of the communication.

Many philosophers have also made the unjustified assumption that somehow written statements can be considered to be something more than communications and can be considered to be either 'true' or 'false' based entirely upon the arrangement of the words. They did this without specifying any process by which a statement can be labelled as 'true' of 'false'. Instead they compounded the error by claiming that truth can be a property of a statement, yet again without denoting any process for how this property can be discerned other than perhaps the rather vague suggestion that this statement is 'true' if it has a 'correspondence with reality'.

Perhaps it is not so surprising that this error was made, for philosophy is a discipline of communication and language is the medium for that communication. However by focusing on language as being something more than mere communication and that somehow words have a direct correspondence with the real world generates numerous so-called paradoxes such as the ship of Theseus. (The Ship of Theseus is the ship that gets repaired over the years to the extent that not one nail or plank of the original ship remains. The dilemma is then whether this is the 'same' ship or a 'different' one.) TPP does not encounter these so-called paradoxes as it considers that words are nothing more than labels and the decision whether to label something as 'A' or 'B' is often arbitrary and insignificant; what is important is the communication. And so it is with the ship of Theseus, it can be labelled as the same ship or as a different ship, it really makes no difference.

From the perspective of TPP, this approach is superseded. According to TPP most thinking takes place below the level of language. The processes involved are those of finding the best pattern to fit the data in order to create a model of the world. Those patterns which are considered indubitable can be labelled as 'true'. Then for the purposes of communication many of those patterns can be mapped onto words that can be spoken or written', in which case those statements which describe an indubitable pattern, according to the communicator, can be labelled as 'true'. The only way that a receiver of the communication can determine whether a statement regarding the real world can be labelled as 'true' for themselves, is to compare the theory described by the statement with the relevant data. This process can only be carried out by a brain with a pyramid of patterns. The logical processes involved are unique to the particular person and also are hidden, they cannot be rigorously and overtly followed by anyone else. There are even hidden from the person themselves. There is no overt and logical process that can demonstrate that a statement about the world is 'true'. In this way statements regarding the real world are necessarily subjective; any claim to objectivity can only be justified on the grounds of consensus.

This description regarding statements refers only to those which are referenced to the real world. Statements which are made regarding abstract systems such as mathematics are somewhat different. This is because the logical processes by which a conclusion is reached are entirely explicit and overt. So the statement '5x20 is 100' can be shown to follow from the axioms of mathematics, hence such a statement can be said to be objectively true, albeit only within the system of mathematics.

Any claim to truth by a philosopher or other person is best treated as a communication and one motivated by the communicator's desire to maximise their own happiness. At best, it will be a subjective truth that upon examination can be labelled as 'true' by the receiver; at worst it can be dismissed as an attempt at propaganda. For in the domain of communication there is a lot of propaganda; and people advocating truths and ways to behave that are designed to benefit the communicator and not the receiver.

That said it is obviously very hard and time consuming to determine for oneself the accuracy of all statements. So it is a meet strategy to accept, at least in the first instance, truths that are true by consensus. These one might find in an encyclopaedia. Such statements are generally uncontroversial and agreed by consensus to be true by most people who have studied such areas of knowledge. Such volumes are of huge benefit to those readers who would otherwise have no knowledge of such matters. This is not to say that everything in an encyclopaedia can be considered to be true by all people but it is certainly a good foundation for an effective model of the world.

It was Aristotle who claimed that logic was at the heart of philosophy and it seems that he was right. Without logic there can be no creation of a model of the world, no form can be extracted from the cosmos.

But what is logic? In its purest form it is the manipulation of symbols according to specified rules. Typically such a process will generate theorems. In TPP there are two main classes of logical systems. There are those that are connected to the outside world via some sort of sensory device such as an eye and which use a pattern identification logic which enables a model of the world to be created. These are called 'Real systems'. Then there are those which have no connection to any exterior world such as pure mathematics, which are called 'Abstract systems'. Both of these classes of systems can lead to the generation of theorems. In the first instance these are patterns which are the best fit to the available data and in the second mathematical theorems which are derived following the logic of mathematics. In both cases the theorems can be considered to be true, albeit only within the system within which they were derived.

'Truth' is used as a label to indicate that a particular idea, theory or statement is acceptable to be a part of one's pyramid of patterns. It is only through logical systems that some form of truth can be obtained. Other claims to truth can be dismissed as not fulfilling the requirements of logical truth. (This is in alignment with Hume's famous statement: "If we take in our hand any book, let us ask: Does it contain any abstract reasoning concerning quantity or number? No. Does it contain any experimental reasoning concerning matter of fact and existence? No. Commit it then to the flames: for it can contain nothing but sophistry and illusion.")

Many philosophers have tried to apply logic to words and language to arrive at what they call truth. However their logic fails to meet the requirements of either a real system or an abstract system. The problem they encounter is that words, or at least the patterns for which they are labels, cannot be treated as simple logical entities or symbols. The relationship between the patterns is very complex and can only be logically processed at the level of patterns, i.e. by a brain with a pyramid of patterns, and not at the level of words.

The only way to process words in a logical system is to treat them as strings of alphanumeric characters (or possibly as sounds but alphanumeric characters are much simpler). By treating them as alphanumeric characters one could also incorporate the entire English dictionary as axioms of a logical system. This would then allow the relationships between the words, albeit not the patterns for which they are labels, to be specified within an abstract system which could then go on to generate theorems. This could be done by cross-referencing the various entries in the dictionary and incorporating rules of syntax and of how to manipulate the strings of alpha numeric characters. These theorems could be of the form 'all sheep are mammals' as the relationship between 'sheep' and 'mammals' can be specified in the dictionary albeit without any links or knowledge of what a sheep or a mammal actually is. This theory could then be labelled as being true within that abstract system. And this theorem would also make sense when processed by a brain with a pyramid of patterns as the relationship between the patterns relating to sheep and mammals is also of the form 'all sheep are mammals'.

However such an abstract system might also generate theorems which make no sense when compared to their meanings within a pyramid of patterns. For example in an English dictionary 'lame' is associated with 'walks with difficulty', 'duck' is associated with 'a batsmen's score of nought' and 'lame duck' is associated with 'an unsuccessful person'. So such an abstract system could take an identity of the form 'A lame duck is a lame duck', switch it around so that it takes the form: 'A lame duck is a duck that is lame' and using simple substitution generate a theorem of the form 'An unsuccessful person is a batsman's score of nought that walks with difficulty'. While such a theorem would be true within that abstract system which deals with words purely as alphanumeric characters it makes no sense when applied to the patterns in a person's pyramid of patterns.

It is only those of us with a pyramid of patterns in our logical processors that can identify the first example as making sense in a real world and the second as nonsense. That is because our brains are able to run logical systems that have access to the patterns that lie beneath the words or labels.

However, many philosophers conflate these systems and claim that the statement 'There are no married bachelors' is 'a priori true' and use it as an example of how some statements can be perfectly and objectively true. However there are a number of logical errors to this claim. If it is claimed to be deduced logically without reference to the world of experience and reality, then it can only be generated by a logical system with access to definitions in the English dictionary as described above. But in that case it is only true within that limited logical system. If it is to be applied to the real world of patterns then there needs to be a mapping between the strings of characters in the statement and the labels the patterns in the pyramid of patterns; and such a mapping is not unambiguous. There is no guarantee that what was true in the abstract system that included the contents of an English dictionary will also be true in the pyramid of patterns as illustrated in the example above regarding the lame duck.

I have described the lack of logical rigour in the claim 'There are no married bachelors' as it illustrates a more general aspect that is useful in making sense of the world; namely that what might seem obvious, simple and logical at first glance may in fact be found to lack logical rigour on closer examination.

There is also a problem in mainstream philosophy with a statement of the form: 'This statement is false' as they cannot decide whether this statement is true or false or neither or both. However according to TPP statements are just another form of communication and are neither true nor false until they are labelled as such. Even the simplest statement: 'This statement is true' is only 'true' if someone labels it as being true because they consider it to be the best pattern to fit the relevant data, though what data that this might be is obscure, or alternatively they consider it to be a theorem generated by an abstract system in which case it is only true within that abstract system. If the statement fits neither of these categories then it is merely a communication, though quite what such a statement indicates can be hard to interpret or perhaps it is nothing more than a random arrangement of words without any particular meaning.

It is important in the process of generating a well-structured pyramid of patterns that one does not include ideas or theories in it that are demonstrably false or do not fit the facts or are so fantastical as to be irrelevant, for such ideas cannot be matched up harmoniously with other ideas because they do not make sense.

The pyramid of patterns that is constructed constitutes a model of the world. And that model is all that one knows about the world.

It may be tempting to consider that one's model is a perfect model of the world and that the patterns that one has constructed from the data are perfectly true. However it must also be recognised that those patterns are only the best patterns to fit the relevant data. It is quite possible that a better pattern could be found at a later date when more data has been accumulated or a better theory tested.

This happens all the time as children grow up and are educated. First of all a child, on being taken to the local duck pond, might 'learn' that all waterfowl are 'ducks'. Later on she might learn that there are also swans and geese, so her interpretation of the data has evolved. What was a very broad pattern of identifying all waterfowl as 'ducks' has now progressed into the division of waterfowl into 'ducks, 'geese' and 'swans'. Later on she might learn to distinguish the gender of ducks and also different subspecies. And there is no reason to think that the development of patterns regarding ducks will end there, for later on she might learn to distinguish and identify individual ducks.

Learning is an ongoing process and the meanings of words are only approximate and they can change as the patterns identified with those words evolve and develop. So if she's makes a statement: "I saw three ducks today", its meaning would be different at different stages of her education.

People want to make sense of the world, they want to take their personal experiences and combine them with the facts that they learn about the world from other people and put it all into one coherent pyramid of patterns.

A coherent pyramid of patterns is conducive to evaluating consequences of possible actions; something that is much harder to do if there are schisms or unbroachable discontinuities between sets of patterns.

Most sets of data are complex and require secondary and tertiary patterns in order to fit the data effectively. For example, there is a common conception that people predominantly act altruistically and only in error do they act in their own self-interest. However this is a conception that is based more on idealism than on the facts. Some people have been led to believe or just want to believe that there is a predominance of altruism in the world and so they hold that pattern as being primary and limit their data points to those that corroborate their original theory. In reality however, and considering all the data points, including those of commercialism, war and crime, it is clear that the fundamental pattern of people acting from self-interest is predominant. People act in order to maximise their own personal happiness. Using this as a primary pattern one can then impose a secondary pattern, over the same data set, that sometimes people will act in what appears to be an altruistic way. For indubitably a person's personal happiness is closely linked to the happiness of other people and so helping others to be happy will have a flow on effect to the person themselves. Also some people want to be seen to act altruistically as they perceive this as benefiting their social status and hence benefiting their self-esteem and how they feel about themselves and hence their own personal happiness.

The idea that people act primarily in an altruistic way simply makes no sense. If someone creates that as the primary pattern, it can only be based on false, biased or incomplete data.

The 14th Dalai Lama put it well when he said: 'I believe that the very purpose of life is to seek happiness'.

And Dale Carnegie put it well when he wrote: 'There is only one way under high heaven to get anybody to do anything. And that is by making the other person want to do it, there is no other way.' This can come from social pressure, financial pressure or perhaps physical pressure. Or to put it another way, people don't do what they don't want to do. Instead people do what they want and that includes politicians, theologians, businessmen, scientists and philosophers.

If this means communicating things that they don't actually believe are true then this is what they will do, just so long as they think that this will bring them happiness and that the repercussions from expressing falsehoods will not catch up with them within their own lifetimes.

The anchor point, which was identified earlier, that happiness is the aim of the brain follows from the logical processes of the brain; it needs a goal in order to function effectively and it seems natural to link this aim with a common word: 'happiness'. Yet this aim is fairly deep and it may not always be clear that this is indeed the aim of a brain. For it may be that at an early stage in a person's life they made a decision, perhaps subconsciously, that the best way to achieve happiness was to follow a particular path which might be, for example, to follow the mores of and support their community, their family, their god or their tradition. Subsequently, their decisions will be in line with this and they will consider that this is their aim in life rather than the pursuit of their own personal happiness. In this way a person may not realise that the goal of happiness underpins all their decisions and is the prime motive in their decision-making.

It then follows that if a person denies that their fundamental goal is to achieve happiness then either they are not aware that happiness is their goal or possibly they simply wish to deny it as they consider that this denial will bring them greater happiness than its admission.

TPP also specifies processes by which other people's theories can be evaluated. Regarding matters to do with the world, the only process that can lead to any theory that has the possibility of being labelled as 'true' is that of pattern identification. If a theory cannot be identified as being the best available pattern to fit the relevant data, it can be discarded. If the data is so scant that no theory can be claimed to be a significant fit, then again the theory can be discarded or alternatively left as a possibility awaiting further data and investigation. This is the way that science works. The best patterns that fit the data are accepted or at least accepted until such time as a better pattern is found.

This same process can be applied to all knowledge or purported knowledge that relates to the real world. It is the way to make sense of the world: accept those patterns that fit the data and reject those that do not. It can be applied to politics, religion and philosophy as well as to science.

While some philosophers might insist that causality and time are a priori or objective entities that are independent of perception and sense data, they are entirely unable to justify this in any sort of logical way. In contrast they are entirely justifiable terms in the TPP paradigm, for they are patterns that fit the available data. Even the concept of existence itself is a pattern that fits the data. These concepts lie very deep in the pyramid of patterns being derived directly from sense data and being at the base level of the pyramid of patterns. They lie so deep that they are fundamental to the whole edifice of the pyramid of patterns that is built above them.

All we know of the world is a model of the world and it is encoded into a pyramid of patterns that we can use to make effective decisions and achieving happiness. While the model may consist of a lot of facts, it is the relationship between those facts that is of vital importance. What is required is for there to be an overall framework for arranging the patterns in a meaningful configuration that enables a smooth transition from one concept to another.

An example of this type of framework is that of the periodic table in chemistry, which upon its development enabled scientific sense to be made of the chemical reactions that are well known previously from the art of alchemy. In effect, the periodic table presents the framework of patterns which facilitated deeper understanding of what atoms and molecules are and how they relate to each other.

TPP aspires to do something similar in the domain of philosophy; to provide a framework for seemingly disparate philosophical ideas which can be related to each other and to achieve a better understanding of the world as a consequence of that. If a suggestion or idea does not fit into this framework then it can be treated with suspicion and either tagged for further investigation or discarded altogether.

As mentioned above, if one is faced with a particular suggestion that purports to relate to a theory about the real world one can evaluate the pattern it suggests by comparing it to the relevant data to determine whether it is the best pattern to fit the data. It is important for this process that all the relevant data is accessed and not merely a few 'cherry picked' data points. If the suggested theory does constitute the best pattern, then it can be accepted, if not, then not.

Should a suggestion refer to a theory regarding some abstract logical system, then it should be possible that once the axioms and processes of inference for that logical system have been identified, for the logical processes to be followed to arrive at a theorem that is equivalent to the original theory. If this is possible then the theory can be accepted, but if not, then not.

Should a suggestion refer to some possible line of action, then those actions can be compared with other possible actions in order to determine whether it seems likely that those actions will maximise one's happiness and minimise one's pain. If it does so then those actions can be taken, but if not, then not.

On the other hand, if one is making a suggestion to another person with the intention of influencing their beliefs or actions, then it is meet to do so in such a way that the receiver of the suggestion will want to hold that belief or take that action as they perceive that it will maximise their happiness and minimise their pain.

If a communication fits none of these categories, then it resides in the domain of fantasy, possibility, play and non-realism. While some of this may be nothing more than idle speculation it may nevertheless serve some useful purpose. Speculation, fantasies, hypothetical situations and playing games can be a productive way of generating possible templates that can subsequently be used in a pattern identifying process that can be applied to data relating to the real world.

Similarly playing games can be a means of identifying possible courses of action together with their possible consequences which can subsequently be used for making real decisions relating to the real world.

There are two primary purposes of the active mind: 1. To create an accurate and efficient model of the world. 2. To interact effectively and efficiently with the world. Communication with others can benefit in both improving one's model of the world and in making better decisions in one's interaction with the world. Communication also allows the learning about other people: who they are, how they view the world and how they interact with the world.

Perhaps I should emphasise that I am making very broad brush strokes in order to describe the TPP framework and how it can be applied to make better sense of the world. In making broad brush strokes I'm hoping to create a model that is applicable to most if not all people. If I went into fine detail it might become more detailed but then only apply to some people. It is in effect a description of a primary pattern; subsidiary patterns will no doubt be needed to actually describe the logical processes of a particular mind. The brain is a highly complex thinking organ that reached its complexity and efficiency through a process of evolution rather than design. It is a remarkable achievement of organic evolution, but it cannot be expected that it is in any way perfect.

One area where the brain would seem to encounter difficulties is that of estimating the probabilities of the possible outcomes of a particular action in order to compare their estimated likelihood of happiness and pain. This takes not only an accurate model of the world but also a lot of computation time in order to assess all the long term consequences.

One way to alleviate the problem is through the use of mores and morals; these are guides to actions that are designed to facilitate decision making and the assessment of long term consequences. Morals are typically biased towards the benefit of the community rather than the individual but nevertheless can be a useful guide.

Some philosophers have tried to assert that ethics and morals are grounded upon some ideal foundation. But they have never managed to identify any such foundation that is indubitable. Morals are better regarded as guides to decision making that have a particular emphasis on the benefits to the community.

While most of the discussion in this book refers to the real world and the identification of patterns from sense data, I have also included some discussion of abstract systems in order to explain how system such as mathematics fit into the TPP framework. Not all logical systems start from sense data. Logical systems can begin with hypothesised axioms and processes of inference that are quite distinct from the pattern identification process. These have no direct link to the real world and can be categorised as 'abstract systems'.

The axioms and processes of an abstract system can be anything anyone wants. But unless they are selected carefully they will be unlikely to generate any interesting theorems or perhaps not generate any theorems at all.

Mathematics is a good example of an abstract system that does generate interesting theories. But this does not mean that any of mathematics is 'true' in a general sense as some mathematicians or philosophers would like to claim. Instead when the label of 'true' is applied to the axioms or theorems of mathematics, they can only be considered to be 'true' within the system of mathematics. The theorems can be labelled as 'true' to indicate that they are part of the mathematical system and the axioms can be labelled as 'true' as they are often indistinguishable from the theorems particularly in a system such as mathematics for which the axioms are implied rather than explicit.

Of course there is conflation between the elements of mathematics and patterns in the pyramid of patterns; the mathematical element '2' is inexorably linked to the pattern labelled as 'two', but in effect this is a mapping that has been exhaustively tried and tested and been found to be useful.

Although mathematics is the paramount abstract system there is nothing magical about it. The fact that most of physics can be effectively modelled using mathematics does not lead to a conclusion that the physical world is defined by mathematics. Physics is no more than a model of the physical world and requires a mapping between measurements of physical entities and the elements of the mathematical abstract system before it can be shown that mathematics can be effective in generating or identifying physical laws.

Mainstream philosophy, from the time of Socrates and his investigation into the meaning of words such as 'beauty' and 'good', has been focused on language or more specifically the meaning of words. It has been assumed that words have a direct connection with the real world. However in order for this to work it requires an assumption of what is called 'naive reality', which means that the world is exactly as we perceive it. It is then further presumed that words represent objects in the real world.

However, this is a naive assumption and has no more justification than the assumption that the world is flat. Both emanate from a simplistic view of the world and can only be justified if there is a limited amount of data available or perhaps a limited imagination as to the possibilities of making sense of the data. In the larger picture both assumptions create anomalies that cannot be resolved.

TPP takes a different approach. It takes a justified view that knowledge of the world is based upon identifying patterns in the data and then labelling those patterns with words. It then follows that words do not so much have meaning as that they are labels for ideas or patterns. It then follows that language has the prime purpose of communication between people. It also follows that knowledge about the real world cannot be encapsulated in the arrangement of those words in statements.

Much of mainstream philosophy focuses on individual 'facts' and the claim that somehow these 'facts' are 'true', though it is more than a little vague about the process by which these 'facts' are created and by what process it is determined whether a 'fact' is 'true' or not.

In contrast TPP focuses on 'understanding' which is the linking of 'facts' together into a cohesive and comprehensive framework. These 'facts' are generated through a process of pattern identification and while some of these 'facts' may be labelled as 'true' to indicate that it is considered that they are the best patterns possible, it is also accepted that it is possible that upon further investigation these 'facts' may need to be amended.

Truth relates to certainty and certainty is much simpler than doubt. With certainty one can simply have or store 'A is true', but with doubt one has something more like 'probably A but possibly B, and C cannot be entirely excluded'. However as all knowledge can only relate to one's model of the world, the certainty that one might ascribe to the world can only apply to one's model of the world.

TPP is a paradigm, a paradigm of simplicity and clarity and one that is highly effective at modelling the world. Part of the process of accepting a new paradigm is being aware of the problems of other paradigms. This is why I have gone to some lengths to point out the logical inconsistencies and inadequacies at the core of the philosophy that so many mainstream philosophers hold.

So then we come to the question of how TPP makes better sense of the world than the more traditional approach to philosophy.

1. TPP starts from the most elementary logic. It is the logic of the manipulation of symbols according to specific rules and identifying patterns. It shows how knowledge can be created from non-knowledge. Traditional philosophy does not achieve this.

2. TPP links many disparate philosophical ideas into one cohesive framework. It links sense data to decisions to happiness to morality. In traditional philosophy such domains are treated in isolation with no threads connecting one to the others.

3. TPP is explicit about what assumptions it makes and has identified them as 'anchor points'. Assumptions are essential for any philosophical synthesis and it is important that they are clearly identified.

4. TPP is entirely self-supporting. While it builds upon the work of philosophers such as Hume and Kuhn, it does not rely upon the accuracy of their work or use their stature to justify any anchor points.

5. TPP is written in plain English without recourse to obscure or invented words. That said, it does use some words with specific meanings and these are listed in the glossary. Ideas, in their essential state, are simple and so there is no justification for convoluted ideas that require complex words and convoluted arguments.

6. TPP has explicit processes for identifying what can be labelled as true that do not rely on vague assertions of what 'must be true' as is the case with mainstream philosophy. What can be labelled as true are typically patterns, statements and theorems.

   A pattern or theory can be labelled as true if it is considered to be the best pattern possible to fit the available data. ('Best' pattern in this respect is one that is most accurate, and most simple. One that can both model the data and also recreate the data most efficiently.)

A statement can be labelled as true if the arrangement of the words and their associated patterns is consonant with the arrangement of patterns in one's pyramids of patterns.

A theorem can be labelled as true if it follows logically from the axioms and processes of inference of an abstract system, though it should be noted that this truth only applies within the particular abstract system that generated the theorem.

7. TPP is descriptive rather than prescriptive. It does not say what people should do or should think; instead it provides a description of how people do think and how they make decisions about what they choose to do.

8. TPP is a synthesis rather than an analysis. It works outwards from clearly identified anchor points.

9. TPP is internally self-consistent. The logical process of finding the best pattern to fit the data that is at the very core of perception is also used throughout TPP to find the best patterns. The patterns it has identified as anchor points are simple and far reaching.

10. TPP fits the facts, so far as it goes. TPP explains how it is possible for a concept to be created out of nothing more than sense data and a logical processor; something that no other philosophical paradigm has achieved.

There are also some interesting points which emerge unexpectedly from the TPP theory and which tally with the facts.

One follows from the requirement for a brain to look for higher level patterns in its pyramid of patterns at a time which would be most efficient when the brain is not being flooded with sense-data that needs real-time processing. This tallies with sleep, a seemingly essential requirement for all higher order animals.

Another is that in the process of creating higher level patterns, a brain may well identify a pattern that relates to its own existence. It might label such a pattern as 'me' or 'I am' or 'I exist'. In this way a degree of self-awareness would be created.

Following this, TPP makes the prediction that self-awareness can only be created at the higher levels of a pyramid of patterns. It then follows that no artificial intelligence, whether a computer or a robot or some other logical processing device can achieve any degree of self-awareness or consciousness without a comprehensive pyramid of patterns.

It is for these reasons that The Pattern Paradigm makes better sense of the world.

As previously mentioned, philosophy is a paradigm; it creates an overt model of the world.

TPP is fundamentally different and distinct from what most philosophers consider to be philosophy. Nevertheless, there is a choice: one can choose one or the other or even create one's own philosophy.

Which philosophical paradigm one chooses will depend on one's own personal situation and character. Some people are quite happy with the mainstream paradigm and even if they encounter things that they are not entirely happy with, the effort of learning a different paradigm is not commensurate with the happiness benefit that they consider they might achieve by switching.

However for other people, the mainstream philosophical paradigm does not make sense and seems to cause more problems than it solves. It might appear to them that mainstream philosophy tells them that they are an elephant when in fact inside they feel that they are a giraffe. For these people, switching their philosophical paradigm is worth all the effort required to learn a new one as the end result is that it allows them to be, and explain how, they are a giraffe if that is what they felt themselves to be all along.

This book has described a model of the world and the processes by which that model was created; primarily through pattern identification. It describes how a person can logically arrive at the concept of self-awareness and how a person's prime directive is to achieve happiness. It is a comprehensive model but it says nothing about the actual experience of self-awareness nor the experience of happiness. No matter what model of the world one might have, whether it is the paradigm described in this book or some other paradigm, nothing can take anything away from knowing the undeniable certainty that we live in a wondrous universe that has produced the bounteous joy of life and of being alive and being aware of all these amazing things.

The purpose of this book is to enable people to have a better model of the world so that they can gain a better understanding of this wondrous life and hence to make better decisions and to better enjoy their lives.

In this book I have tried to describe the processes by which a basic model of the world can be constructed. I consider that these ideas constitute the blueprint for the best model of the world currently available; best in terms of simplicity, accuracy, depth and comprehensiveness.

The philosophy I have described in this book is intended to be applicable to all people. I have tried to avoid those aspects of philosophy which are specific to a particular culture or a particular people. To this end I have only described the main core of the pattern paradigm philosophy. If philosophy were a tree, then what I have described constitutes the trunk and main branches of the tree. I leave it to the reader with their own particular character, culture and experiences to add the colour and texture to the main ideas that I have described; or following the analogy of the tree, I leave it to the reader to add the smaller branches, twigs, leaves and flowers.

I have found the ideas in this book to be extremely interesting and I've had a lot of fun in exploring them. I invite you to do the same.

Glossary

Words can have many meanings and some words in this book have been used with the intention of a specific meaning. This glossary is an indication of those intended meanings.

Abstract	Has no direct connection to the real world
Algorithm	Sequence of instructions or a sequence of rules to be applied.
Anchor Point	A pattern or idea that is held to be true and which is also held to be particularly important and significant; it is then used to create higher level patterns. They have a similar role to axioms but refer to the real world rather than to an abstract system.
Axiom	Starting point for a logical system
Belief	A pattern that has been identified as being the best possible pattern to fit the relevant data
Brain	The body's logical processor and decision making organ
Cognition	Higher level pattern identification
Conflation	The combination of two different systems. The word is often used to indicate a combination that is not logically justified and can lead to erroneous conclusions

Domain	Range of data points used for searching for a particular pattern or for discussion
Evolution	The process by which life has emerged from inanimate matter
Fact	Pattern that is a good fit to the data
Genotype	Genetic constitution of an organism
Happiness	That to which a brain or mind aspires
Inference	Output or conclusion of a logical system
Interesting	Something that stimulates brain activity, particularly those ideas that have the potential to be applied to decision making
Language	Means by which ideas or patterns can be communicated from one entity to another
Logic	A means of making inferences following specific rules
Logical Process	Logical method by which inferences are made
Mapping	Linkage between the elements of one system with the elements of another. For example the points on the surface of the Earth can be mapped onto points on the pages of an atlas
Mind	Brain activity that relates to the upper part of the pyramid of patterns that is above the level of the pattern 'I am'
Model	Logical representation of something else
Normative	What is considered to be normal or popular
Pain	Opposite or absence of happiness
Paradigm	A set of beliefs that are self-consistent and which describe the world
Parameter	A quantity that is fixed for a particular set of calculations but can vary from one set of calculations to another
Pattern	A compressed form of data. So for example, given data of the form, 1,3,5,7,9 ... the pattern would be '2n+1'
Perception	Process of making sense of raw sense data through pattern identification

Phenotype	Physiological features of an organism
Propaganda	Attempt by one person to influence the thoughts, decisions or actions of another through communication
Pyramid of Patterns	Hierarchy of patterns stored in a brain with patterns created from raw sense data at the bottom and each higher level generated by identifying patterns created from the patterns below.
Real	Is connected to the real world through sense data
Real world	That which is presumed to exist beyond sense data
Recursive	The process by which the results of a logical process are subsequently used as input for the same logical process
System	A logical unit that is capable of making inferences
Template	Initial seed used in the search for a pattern
Theorem	The conclusion of some processes of a logical system
TPP	Acronym for 'The Pattern Paradigm', the philosophical paradigm introduced in the book of the same name and continued in this book.
True	Label attached to a belief to indicate confidence that it is the best possible belief; particularly used in communication
Understanding	Having a framework that links facts together in a cohesive and comprehensive way
Word	Label for a pattern; usually one that is capable of being spoken or written in a communication

ADDENDUM

If you would like to comment on or discuss the ideas in this book, then please visit: www.patternphilosophyforum.com

www.ingramcontent.com/pod-product-compliance
Lightning Source LLC
Chambersburg PA
CBHW030812180526
45163CB00003B/1254